Daniel Vella, Mathias Fuchs (eds.)
Digital Culture & Society (DCS)

Editorial

Digital Culture & Society is a refereed, international journal, fostering discussion about the ways in which digital technologies, platforms and applications reconfigure daily lives and practices. It offers a forum for critical analysis and inquiries into digital media theory. The journal provides a publication environment for interdisciplinary research approaches, contemporary theory developments and methodological innovation in digital media studies. It invites reflection on how culture unfolds through the use of digital technology, and how it conversely influences the development of digital technology itself.

Board:
Maria Bakardjeva (University of Calgary), David Berry (University of Sussex), Jean Burgess (Queensland University of Technology, Brisbane, Australia), Mark Coté (King's College London), Colin Cremin (University of Auckland), Sean Cubitt (Goldsmiths, University of London), Mark Deuze (University of Amsterdam), José van Dijck (Utrecht University), Delia Dumitrica (Erasmus University Rotterdam), Astrid Ensslin (University of Alberta, Edmonton), Sonia Fizek (Abertay University), Federica Frabetti (University of Oxford), Orit Halpern (The New School, New York), Irina Kaldrack (Braunschweig University of Art / Leuphana University of Lüneburg), Denisa Kera (National University of Singapore), Lev Manovich (The Graduate Center, The City University of New York), Janet H. Murray (Georgia Institute of Technology, Atlanta), Jussi Parikka (University of Southampton), Lisa Parks (University of California, Santa Barbara), Dominic Pettman (The New School, New York), Rita Raley (University of California, Santa Barbara), Richard Rogers (University of Amsterdam), Julian Rohrhuber (Robert Schumann School of Music and Media, Düsseldorf), Marie-Laure Ryan (University of Colorado, Boulder), Mirko Tobias Schäfer (Utrecht University), Jens Schröter (University of Bonn), Trebor Scholz (The New School, New York), Tamar Sharon (Maastricht University), Roberto Simanowski (City University of Hongkong), Nathaniel Takcz (University of Warwick), Geoffrey Winthrop-Young (University of British Columbia, Vancouver), Sally Wyatt (Maastricht University)

Daniel Vella (PhD), born in 1984, teaches game studies at the Institute of Digital Games at the University of Malta. He read for his PhD at the IT University of Copenhagen, focusing on the development of a theory of ludic subjectivity. His research interests include the phenomenology of virtual world experience, place and architecture in games and aesthetics in digital games. He regularly publishes in journals such as Game Studies, the Journal of Virtual Worlds Research and Techne': Research in Philosophy and Technology.

Mathias Fuchs (Dr.) is an artist, musician and media scholar. He is the director of the Gamification Lab at Leuphana University in Lüneburg. He is a pioneer in the field of game art and is a leading scholar in game studies and directs a project on Gamification that is funded by the German Research Council (2018-2021).

Daniel Vella, Mathias Fuchs (eds.)
Digital Culture & Society (DCS)
Vol. 10, Issue 2/2024 – Ruins and the Contemporary

[transcript]

Bibliographic information published by the Deutsche Nationalbibliothek
The Deutsche Nationalbibliothek lists this publication in the Deutsche Nationalbibliografie; detailed bibliographic data are available in the Internet at https://dnb.dnb.de

Indexed in EBSCOhost databases.

© 2025 transcript Verlag, Bielefeld
transcript Verlag | Hermannstraße 26 | D-33602 Bielefeld | live@transcript-verlag.de

All rights reserved. No part of this book may be reprinted or reproduced or utilized in any form or by any electronic, mechanical, or other means, now known or hereafter invented, including photocopying and recording, or in any information storage or retrieval system, without permission in writing from the publisher.

Printed by: Elanders Waiblingen GmbH, Germany
Cover layout: Kordula Röckenhaus, Bielefeld
Typeset: Mark-Sebastian Schneider
Print-ISBN 978-3-8376-7320-3
PDF-ISBN 978-3-8394-7320-7
ISSN of journal: 2364-2114
eISSN of journal: 2364-2114

Printed on permanent acid-free text paper.

Content

Ruins and the Contemporary
Introduction
Mathias Fuchs and Daniel Vella   7

I   Ruins as Idea

Videogames, Wastelands and War
Rethinking Possibility Spaces
Souvik Mukherjee   17

Gaming in Ruins
Alienation and Hope in *NieR: Automata*
Timothy Welsh   33

"This Whole Place is Built from Ghosts"
Playfully Imagining Community and Hope
in the Ruins of Capitalism in *Citizen Sleeper*
Ian Sturrock   51

Ruin
A Call to Becoming-at-Home Again
Daniele Monaco   69

II   Ruins as Image

Moving Image. Changing Landscape
Trevor Borg   95

Allegorical Ruins and the Possibilities of Human Futures
A Case Study of the Game *Lifeless Planet*
*Caio Tulio Olimpio Pereira da Costa
and Ana Laura Matos Torquato*   111

The Ruins Of Home, Hearth, Kingdom, and Man in *Breaking Bad*
Ruins, Ruination and Ruin Porn in 'The ABQ'
Michael J.T. Stock   131

"A Beachfront Property in Gaza is Not a Dream"
Utopia, the Digital, and the Image of the Ruin in Palestine
Daniel Vella   147

III Ruined Places

A Performance of Decay
Amitesh Grover's Site-Specific Theatre, *The Money Opera*
Rahul Bishnoi   167

Paratextual Anarchaeology
Revisiting *WildStar* Through its Residual Traces
James Manning and Lawrence May   177

Post-Apocalyptic Ruins in Digital Games as Indexical Storytelling Devices
Romi Sofia Abatangelo   203

Biographical Notes   219

Ruins and the Contemporary
Introduction

Mathias Fuchs and Daniel Vella

When we started planning for this special issue of the *Journal of Digital Culture & Society*, the international public was confronted with images of individual East-Ukrainian buildings destroyed by Russian artillery. Little did we know at the time that the small number of houses that became ruins then, would turn into wide fields of massive ruination. More than a year after the Russian invasion of Ukraine, another war zone provided us with images of ruins in the Gaza strip. The number of buildings that have been destroyed beyond repair is enormous. According to satellite images obtained by the Hebrew University Geographic Information System Center, 70 % of residential buildings in Gaza have been destroyed in the last two years (Hasson 2025). In Rafah, 89 % of the city have been "leveled" (Aljazeera 2025). By 2024, the hostilities had left one million people homeless (Goldstein/Staff 2024), and current estimates suggest that some 1.4 million out of 2.3 million inhabitants have lost their homes (Euronews 2025).

In Ukraine, more than 250,000 housing units were destroyed or damaged, including 222,000 private homes, 27,000 multi-family dwellings and 526 dormitories (Kiev School of Economics 2024). According to satellite image evaluations, a quarter of a million buildings have been ruined, including 900 schools, hospitals and other institutions. None of these numbers can be verified, but as both conflicting sides report similar quantities of ruination – either as a loss, or a glorious victory – one can assume that they are roughly correct. A new military strategy seems to have emerged that aims at destroying buildings and civilians – as opposed to killing enemy soldiers.

It is not only the scale of ruination that has changed, but also the sustained and irreversible nature of the destruction. A comparison of press photos from the Six-Day War (June 5 to 10, 1967) and the on-going Gaza hostilities (October 7, 2023 – present) shows images of destroyed tanks and shot-down airplanes in the case of the former, and ruined civil buildings in the case of the latter. Whereas propaganda images and press reports of reconstruction and rehabilitation have always been an integral part of the narrative of ruination in former wars, we see no signs of a reversal of the destruction in the current wars.

The hope for salvation has become feeble. Hartmut Böhme stated in 1989 that the ruin "[...] demonstrates a precarious balance of obtained form and decay, nature and history, violence and peace [...]" (1). Do contemporary ruins still offer the prospect of peace? And how are nature and history balanced out, when aggressive ruination is visited upon the world's vital ecosystems?

Flora and fauna (including human animals) are under fire by military operations. A recent study indicated that the first 120 days of hostilities in Gaza produced more greenhouse gas emissions than 26 individual countries and territories produce in a year (Out-Larbi et al. 2024). They are also under attack through the human destruction of forests, glaciers, coral reefs, and other essential habitats, along with the ever-growing threat of the climate crisis – a great deal of which can be attributed to the extractivist, and colonialist, logic of industrial capitalism. In her 2015 book *The Mushroom at the End of the World: On the Possibility of Life in Capitalist Ruins*, the anthropologist Anna Lowenhaupt Tsing warns:

"When its singular asset can no longer be produced, a place can be abandoned. The timber has been cut, the oil has run out; the plantation soil no longer supports crops. The search for assets resumes elsewhere. Thus, simplification for alienation produces ruins, spaces for abandonment for asset production. Global landscapes today are strewn with this kind of ruin." (Tsing 2015: 6)

This issue of *Digital Culture & Society* addresses the complex thematic field of ruination and decay in contemporary culture. In our lived environments, our media landscapes, and our social media feeds, we are surrounded by ruination: the ruins of war, ecological disaster, post-industrial decline, austerity, social collapse, and infrastructural decline.

The cultural fascination with ruins is nothing new. Architects, poets and commentators like Rose Macaulay (1953), Christopher Woodward (2002) and Robert Ginsberg (2004) have extensively and critically investigated the significations of ruins in Western culture. Salvatore Settis (1997) unpacks the multivalence of the figure of the ruin when he writes that:

"In their persistent presence, ruins speak to us of the structures they once were, of the people who made them, of those who commanded them to be made, and of those who for a time made use of them. In their evocation of absence, they speak of those who destroyed them or abandoned them or failed to protect them from the irresistible ravages of Time. In their present state, ruins speak of those who have tried to make sense of them, or have been drawn to represent them, or have used them as objects of memorialization."

It seems, however, that in the contemporary moment, our relation to ruins takes on new shapes, radically different from that which spurred the Grand Tourists to make their pilgrimages to the sites of Classical antiquity. Ruins have long lost the solemn serenity of a *memento mori* Böhme could still write about in 1989, comparing the aestheticisation of ruins in Western culture from 1337 to the present with an interest in "Memory – Scripture" (Böhme 1989: 287), with "the City of Rome as a focal point for a longing for ruins" and ruins in the "Posthistoire" (302). In our present moment, ruins do not primarily signify the monumental remainder of a lost Classical golden age to which the cultural elite can lay a claim of affinity, nor do they only anchor melancholy Romantic contemplations of loss and transience. Though the games and subsequent TV adaptations of *The Last of Us* (2023) and *Fallout* (2024) imagine the present in ruins, such images speak to us in a different register to Hubert Robert's 1796 painting of the Grand Gallery of the Louvre in ruins. Ruins are no longer simply the object of reflection at a safe distance, or a medium of historical awareness. Rather, in the midst of war, irreversible climate change, and massive economic and ecological disasters, ruins are what surround us. Increasingly, the natural and built environments we inhabit and are familiar with are marked by ruin, and they are present in literature, film, digital games (Vella 2010; Watts 2011; Lowe 2012; Fraser 2016), and in news feeds saturated with images of the aforementioned, present-tense ruins of Ukraine and Gaza.

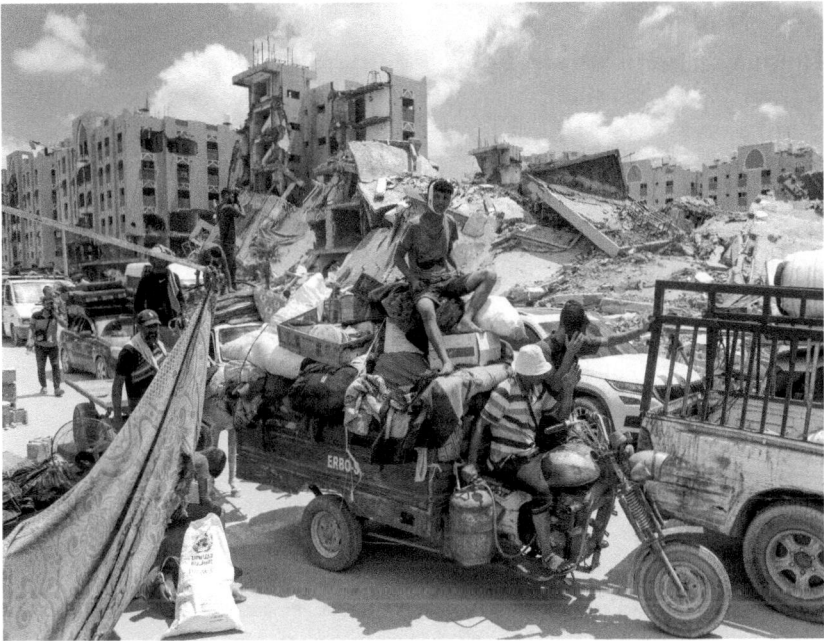

Fig.: Refugees enforced to flee Hamad quarter in Khan Younis, southern Gaza Strip (photo: Ashraf Amra, 2024, CC BY-SA 4.0)

We find ourselves in a society that cannot be completely convinced of a hope for salvation via either technology or politics, and does not want to accept an apocalyptic end of times. Berlin-based techno club "22 kW of Sound" describes the attitude of their clientele in these words:

"We are living in a world, where parts of reality seem dystopian and a few dystopias are already real. We do not close our eyes – yet we will keep dancing and we need to dance."

Such a response to William Gibson's observation that dystopia is not a future possibility but already an unevenly distributed reality invites interrogation. Is this the answer to the problem? Is there a way out apart from dancing? Or have we returned to a seventeenth-century fatalism expressed in the words "[...] non si può guarire, bisogna morire." (We cannot be cured, we all must die!) (Filippo Neri quoted in Färberböck/Mayer 2023: 33).

It is in the spirit of such an interrogation that we invited contributors to consider what to make of the ruin in contemporary culture and society. Robert Ginsberg reminds us that "[t]he poetics of ruin must have room for the politics of ruin" (2004: 144). How do we react to "brand new" ruins as opposed to the ruins from antiquity (Hell/Schönle 2010)? How can we cope with geopolitical processes that create abandoned spaces in one place and crowded hubs in other places (Fraser 2016; Dzenovska 2020)? How shall we respond to the collapse of the infrastructure that we used to take for granted: transport, energy, health services, and the very heating and air-conditioning we need more than ever to survive climate extremes, even as they contribute to worsening the climate crisis? How does the image of the ruin resist the discourses of technocapitalist positivism and neoliberal promise? What are the skills we need to learn to live among the ruins? Do we have to develop a new kind of "rubble literature" (Buchanan 2007) to be able to talk about what is happening on our planet? How does the materiality of the ruin trouble, or resist, the promise of the digital and its reconfiguration of social reality? Can ruins help us imagine "the world without us" (Weisman 2007), and in doing so, let us glimpse a posthumanist vector out of the Anthropocene (or the Capitalocene)?

The ruin, now more than ever, is an inherently polysemous figure. It is no surprise, then, that the contributors to this special issue of the Journal have approached the ruin, and the processes of ruination, along different paths. We have grouped these contributions into three broad thematic sections.

The first section, Ruins as Idea, includes perspectives on ruination as a critical or philosophical concept.

Souvik Mukherjee extends and follows up on former publications on the Wasteland as 'Any-Space-Whatever', a term he borrows from Deleuze (cf. Mukherjee 2010; 2012; 2019; 2020). Still interested in the wasteland and ruins as archetypical settings for digital games, he turns his attention to the affective qualities of wastelands and the wasteland as a metaphor. Mukherjee argues "the wasteland may serve as an illustration of the Deleuzian time-image which can be interpreted as pure potentiality." This makes the wasteland a possible space, or as Massumi has it, "a state of suspense, potentially of disruption."

Timothy Welsh dissects the game *NieR: Automata* (2017), in order to find out how alienation and hope result from gaming experience in ruins. An engagement with the posthuman (and post-Anthropocene) ruinscapes of the gameworld is intertwined with a discussion on the idea of 'gaming in the ruins,' to refer to the perception – prevalent in many parts of gaming culture – that, when we engage with video games in the contemporary moment, we are "subject to and participating in modes of alienation that characterize the end of capitalism."

Ian Sturrock close reads *Citizen Sleeper* (2022), using lenses of empire, capitalism, adaptation, and ruin. Sturrock has no doubt that "energy- and resource-intensive modes of entertainment, intertwined with, and compromised by, the capitalist, extractivist system," frame the gaming experience. He is, however, not without hope, when he sees a certain value "in game designs that allow us to imagine alternative pleasures and satisfactions to those of capitalism, which *Citizen Sleeper* clearly does."

Daniele Monaco, meanwhile, adds an example game of a "natural ruin in an 'artificial' world" to his philosophical investigations. The game is called *Stardew Valley*, and the philosophical proposal starts with a reading of Heidegger's ideas on ruins as the symbol of decaying of life.

The second section, Ruins as Image, focuses upon the mediation and representation of images of ruins and ruination in the contemporary media and cultural landscape.

Trevor Borg discusses *Olea*, a visual, auditory, and olfactory installation. The work features a 'mythical' figure situated within an indeterminate temporal realm that spans across historical epochs. *Olea* was filmed in four different locations in Malta, two of which are directly linked to the Roman period. Featured locations comprise Il-Mixquqa (Golden Bay), a sandy beach in the northern part of Malta not far from a complex of Roman baths, San Pawl Milqi in Burmarrad, where extensive Roman oil production architecture has been uncovered, and the Domus Romana, a museum. The audience was encouraged to virtually 'edit' the sequence, starting and ending the viewing at any point in time to encourage alternate and unlikely readings.

Caio Tulio da Costa and Ana Laura Torquato conduct a case study on the game *Lifeless Planet* (2014). This game, they say, "uses ruins as narrative and experiential elements, positioning them as catalysts for reflection on human ambition and future potential."

Michael Stock focuses upon a particular instance of ruin in the TV series *Breaking Bad*, investigating a flashforward sequence in the episode "Blood Money" from the show's final season (2013) in which the protagonist, Walter White, revisits his family home and finds it abandoned. The sequence is read in the context of the show's vision of ruination as being inescapable in the contemporary American city.

Daniel Vella, one of the editors of this special issue, starts his analysis from an image shared on social media by the Israeli real estate firm Harey Zahav in December 2023, two months into the assault on Gaza, showing a line of wire-framed luxury beachfront villas superimposed upon a photograph of ruined buildings in Gaza. Contrasting the significance of the ruin in relation to utopian discourses – and making the claim that ruins are that which lingers in utopia's shadow – he argues that the image is emblematic of a shift away from utopian imaginings that obscure the violence of their creation.

The third and final section, Ruined Places, includes contributions whose investigations focus on specific sites of ruination, whether in the real world or in virtual spaces.

Rahul Bishnoi analyses Amitesh Grover's experimental performance *The Money Opera*. Staged within the ruins of abandoned buildings in Goa and Delhi, the production invites audiences to navigate a multilayered narrative space. Actors present billionaires, ghosts, poets and stockbrokers as parties of a play that represent contradictions and hidden truths of capitalist society.

James Manning and Lawrence May offer a methodological intervention that allows encounters of play to be reconstructed from the past by drawing on digital ephemera created by players. They call this method an "anarchaeology" and follow up on ideas of media scholar Zielinski (2006). Very much aware of the fact that ruins of formerly trending video games cannot be completely comprehended when the hardware or an emulation of the game is presented in museums or exhibitions, they look at the "paratext" of games and play. This aims at reconstructing a game that is not any longer widely available from its paratextual remnants. Fan culture, packaging, gaming environments are components that need to be observed to get rid of the simplification of a object-cantered preservation of games.

Finally, **Romi Sofia Abatangelo** writes about post-apocalyptic ruins in digital games as indexical storytelling devices. Their references are to the digital games *Horizon Zero Dawn* (2017) and *Disco Elysium* (2019). One of their observations is that both games use ruins to construct "explorable spaces that need active engagement from players to discover and interpret clues gathered from the ruins in order to reconstruct the past history of the game world."

Acknowledgements

This publication evolved from the 2024 conference "Alles bröckelt" (Everything crumbles) that was organized by the International Research Center for Cultural Studies (ifk) in Vienna in cooperation with L-Università ta' Malta. and University of Art and Design Linz. Scholars and artists participating in the conference prepared the ground for what we attempt to theorise about in this issue of the *Journal of Digital Culture & Society*. We would like to thank the director of *ifk* Karin Harrasser and the former director Thomas Macho for their inspiring ideas. We would also like to acknowledge the academic and artistic input of the conference participants Paul Dobraszczyk (Bartlett School of Architecture, London), Dace Dzenovska (University of Oxford), Emma Fraser (University of Berkeley), Orit Halpern (TU Dresden), Julia Hell (University of Michigan), Mina Lahlal (Ärztin), Thomas Macho (Berlin/Vienna), Rosalind C. Morris (Columbia University), Leonhard Müllner (Total Refusal), Omar N'Shea (L-Università ta' Malta), Monika Wagner (Universität Hamburg), Katharina Weinberger-Lootsma (Kunstuniversität Linz) and students of Kunstuniversität Linz.

References

Böhme, Hartmut (1989): Die Ästhetik der Ruinen. In: D. Kamper / Chr. Wulf (ed.): Der Schein des Schönen. Steidl Verlag, Göttingen. pp. 287-304.

Byles, Jeff (2006): Rubble: Unearthing the History of Demolition. Three Rivers Press, 2006.

Dzenovska, Dace (2020): Emptiness: Capitalism without People in the Latvian Countryside. In: American Ethnologist. 47, 1, p. 10-26.

Färberböck, Peter/Mayer, Aska (2023): Non si può guarire: An (idea)historical approach to plague games and death in the streets. In: Apgar, Blair (ed.) The Middles Ages in Modern Games. University of Winchester, pp. 33-37.

Fraser, Emma (2016): Awakening in ruins: The virtual spectacle of the end of the city in video games. In: Journal of Gaming & Virtual Worlds. 8, 2, p. 177-196.

Ginsberg, Robert (2004): The Aesthetics of Ruins. Rodopi, Amsterdam and New York.

Goldstein, T./Staff, T. (2024): In "Times of Israel" January 24, 2024.

Hasson, Nir (2025): "New Satellite Data Shows: Gaza Devastation Scale Greater Than Estimated, at Least 70 Percent of Buildings Leveled" In: Haaretz, July 17, 2025. Available at: https://www.haaretz.com/israel-news/2025-07-17/ty-article-magazine/.premium/satellite-data-shows-at-least-70-percent-of-gaza-buildings-leveled/00000198-12de-d9c7-af98-7adffc8f0000

Hell, Julia/Schönle, Andreas (2010): Ruins of Modernity. Duke University Press.

Kiev School of Economics (2024): 155 billion. The total amount of damages caused in the Ukraine. Available at: https://kse.ua/about-the-school/news/155-billion-the-total-amount-of-damages-caused-to-ukraine-s-infrastructure-due-to-the-war-as-of-january-2024/

Lowe, Dunstan (2012). Always Already Ancient: Ruins in the Virtual World. In: Thorsen, T.S., (ed.) Greek and Roman Games in the Computer Age. Akademika Publishing, Trondheim, Norway, pp. 53-90.

Macaulay, Rose (1953): Pleasure of Ruins. Thames and Hudson.

Mukherjee, Souvik (2010): "'The Water of Life Freely': Water and the Wasteland in Fallout 3". In: *Ludus Ex Machina*, Available at: http://readinggamesandplayingbooks.blogspot.com/2010/01/water-of-life-freely-water-and.html (Accessed: 13 January 2025).

Mukherjee, Souvik (2012): "EgoShooting in Chernobyl: Identity and Subject(s) in the S.T.A.L.K.E.R Games". In: Handbook of Digital Game Cultures. Springer.

Mukherjee, Souvik (2019): "Videogame Wastelands as (Non-)Places and 'Any-Space-Whatevers'". In: Aarseth, Espen/Günzel, Stephan (eds.): *Ludotopia: Spaces, Places and Territories in Computer Games.* Bielefeld: transcript Verlag, pp. 167-184.

Mukherjee, Souvik (2020): "Coming of Age in the Capital Wasteland: The Videogame Narrative as a Space of Possibility". Refractory: a Journal of Entertainment Media.

Out-Larbi, Frederick, Benjamin Neimark, Patrick Bigger, Linsey Cottrell, Reuben Larbi (2024). "A Multitemporal Snapshot of Greenhouse Gas Emissions from the Israel-Gaza Conflict." Available at: https://www.qmul.ac.uk/sbm/media/sbm/documents/Gaza_Carbon_Emissions.pdf [Accessed 4 August 2025]

Settis, Salvatore (1997): Introduction. In Irreversible Decay: Ruins Reclaimed, ed. by Michael S. Roth and others (Los Angeles, CA: The Getty Research Institute for the History of Art and the Humanities.

Tsing, Anna Lowenhaupt (2015): *The Mushroom at the End of the World: On the Possibility of Life in Capitalist Ruins.* Princeton University Press

Vella, Daniel (2010): Virtually in Ruins: The Imagery and Spaces of Ruin in Digital Games.MA dissertation: University of Malta.

Watts, Evan (2011). Ruin, Gender, and Digital Games. In: Women's Studies Quarterly, 39, 3/4, pp. 247-265.

Weisman, Alan (2007): The World without us. Thomas Dunne Books, New York.

Woodward, Christopher (2002): In Ruins: A Journey Through History, Art, and Literature. Knopf Doubleday Publishing Group.

Zielinski, Siegfried (2006): Deep Time of the Media – Towards an Archaeology of Hearing and Seeing by Technical Means. MIT Press, Cambridge MA, London.

Ruins as Idea

Videogames, Wastelands and War
Rethinking Possibility Spaces

Souvik Mukherjee

Abstract

The recent Amazon Prime video series, Fallout, remediates the experience that many gamers the world over have of traversing a wasteland fraught with the imminent (and immanent) sense of possibilities, fear and danger in the Fallout videogames franchise, particularly Fallout 3, Fallout: New Vegas and Fallout 4. Previous discussions of the game have addressed how the wasteland functions as a metaphor in many videogames and is an affective space, and to use a phrase used by Gilles Deleuze, is an espace quelconque or 'any-space-whatever'. Deleuze introduced this concept in his Cinema books (1983; 1985) and used it for as disparate scenarios as the empty spaces in the films of Yasujiro Ozu and a railway station in Paris in Robert Bresson's The Pickpocket. Both examples, however, are spaces that are throbbing with possibilities that are awaiting actualisation. Arguably, this is also true of videogame spaces and games such as the Fallout series, the Borderland series, the Stalker series and other videogames that are directly set in a wasteland.

Extending the wasteland metaphor to other videogames and videogame genres (particularly real-time strategy games, which involve construction, planning and resource management), this article aims at exploring the boundaries of how and where the any-space-whatever concept can be applicable in videogames, particularly those that aim at narrative multiplicity and challenge any fixed telos. Especially relevant in the current political contexts where cities are being changed into 'wastelands', one question arises anew. To echo T.S. Eliot's celebrated poem's question:

What are the roots that clutch, what branches grow
Out of this stony rubbish?

Videogames portraying war-ravaged spaces and ruins turn their locations into affective spaces and arguably give them new relevance and meaning in viewing them as throbbing with possibilities that shape the identities of the people (and non-human lifeforms) that inhabit them. This article connects the wasteland metaphor from games such as Fallout (and their remediations in television and print) to other larger usages of the wasteland as a metaphor and then examines how videogames portray different kinds of wastelands, such as the war-torn backdrop of This War of Mine *or* Bury Me, My Love, *and create possibility spaces that reshape identities and the very idea of what it is to be human under circumstances of war and destruction.*

Keywords

wastelands, affect, war, any-space-whatever, apocalypse

Remediating the Wasteland in Videogames

The Wasteland is one of the most common settings in videogames. It is a *mise en scene* that signifies danger, deprivation, chaos and most often, the concomitant need for restitution. Recently, in the 'over the top' (OTT) productions of online television serials that adapt videogames into television, the wasteland has emerged as a prominent setting, especially in adaptations of videogames such as *Fallout* and *Borderlands* to varying effects as this article will point out. Earlier an article on videogame wastelands had made a connection with cinema and described these spaces as the any-space-whatever, a concept introduced by Gilles Deleuze in 1983 in his *Cinéma 1* where he discusses the films made by Yasuziro Ozu and Robert Bresson as spaces throbbing with possibility. This article extends that definition and views the wasteland as both a metaphor and a mechanic that is essential to understanding videogames. In the way videogames deploy the conception, one needs to view this as a remediation or a 'refashioning' of media as Jay David Bolter and Richard Grusin (2000) describe the process in which older and newer media adapt and/or relate to each other. As such, the connections are not just to be made between games and film but to older media such as literature, for example Thomas Stearns Eliot's (2013 [1922]) famous poem *The Waste Land*. On looking at the videogame wastelands, their remediations of wastelands in earlier media and in turn, their remediations into forms of earlier media (the OTT is both reflecting its antecedents in film and television as well as its own media specificity), one can begin to see how videogames emphasize the importance of wasteland-like spaces is narrative media and the role that they play in the construction of identities, notions of temporality and narrative construction.

The wasteland, it will be argued here, serves as an all-pervading metaphor for videogames. In addition, there are other sub-tropes that play an important part in the shaping of the wasteland as the *mise-en-scene*. In *Fallout 3* (Bethesda 2008), it is water and in *Stalker: Shadow of Chernobyl* (GSC Game World 2007), one could say that it is the location itself, the Zone. These two games will be discussed as typical examples of wastelands in videogames because they are similar post-apocalyptic locales where the very ground is irradiated and often deadly.

The Wasteland as Metaphor: Water and the Zone

As pointed out elsewhere (2010), the key concern with water is not often noted in the scholarship[1] around the game:

The key argument of *Fallout 3* comes from the book of Revelations: 'I am Alpha and Omega, the beginning and the end. I will give unto him that is athirst of the fountain of the water of life freely.' The other and more obvious connection between water and the vast area of the Capital Wasteland (roughly present-day D.C and outlying areas) is, ironically, the absence of water in the stony rubbish reminiscent of Eliot's poem. There is, obviously, a section of the game that features the Potomac river and others that show areas submerged in water. This, however, is 'dirty' water – deadly, irradiated and often the home of dangerous mutants appropriately named Mirelurks. Pure water, although in extremely short supply, is available in small quantities throughout the game and before the Lone Wanderer manages to fulfill his quest, the only source of purified water in the game is seen in the town of Megaton which is fighting a constant battle to maintain its water purifying plant in working condition. Indeed, one of the minor quests that the player finds in Megaton is to repair its leaking water pipes. It is not clear whether the other large human habitations in the game, such as Rivet City (an erstwhile aircraft carrier) and the Citadel have any store of pure water. (Mukherjee 2010).

Fallout 3's concern with water is not new for accounts of wastelands in other media and the game itself points at connections with earlier narrative media, whether directly or indirectly. Again, elsewhere, I point to connections with literary sources such as the nineteenth-century English poet Gerald Manley Hopkins's sonnet 'Thou Art Indeed Just, My Lord' and T.S. Eliot's *The Waste Land*:

Despite the problems of deciding on the way to make the wastelands fertile again, water is still most commonly seen as the solution as perhaps embodied by Gerald Manley Hopkins's famous line, 'God, send my roots rain'. Hopkins is speaking of the wasteland of his mind and he places himself in the tradition followed by many others (including Eliot), where the wasteland is an allegorical device. In a way, *Fallout 3* is an allegory as well as a literal chilling reminder of a possible physical wasteland that might be the result of our failure to keep peace with each other. (ibid)

The game also directly references American poet Sara Teasdale's (2012) poem, 'There Will Come Soft Rains' and Ray Bradbury's (1989) science fiction story by

---

1   Melissa Bianchi's (2024) 'Beyond Barren Wastelands: The Greening of the Post-apocalypse in Video Games' addresses the importance of plants and foliage in game design, particularly to encourage a post-anthropocene thinking in game design but it does not directly address the significance of water as an agent of fertility.

the same name, both works in which the extinction of mankind is addressed. Bradbury's story complicates Teasdale's notion of the indifference of nature to mankind's self-induced extinction: 'The indifferent cycles of nature that Teasdale invokes are refigured by Bradbury in the automated household, as if all of the technological achievements that were intended to insulate human beings from the environment have become just another implacable form of indifference to human well-being' (ibid). In *Fallout 3*, should the player come to the location of the McClellan household, she will come across a 'Mr. Handy' robot from the game that recites Teasdale's poem in a scene that is poignantly reminiscent of Bradbury's story and also reflects the starkness of the post-apocalyptic Capital Wasteland (what remains of Washington D.C.) where many of the original inhabitants have not survived and where the 'soft rains' that come may be the 'dirty' radioactive water. As a player comments, 'Ever wonder why it never rains in the game? It could add a whole new aspect to the game. The rain could be severe acid rain that eats away your armor, and then your health. Also lightning and thunder would be cool to [sic]. I think a thunder storm out on the wastes at night would look sweet' (Wallace 2009). Rains, here, no longer signify fertility in the restorative sense of Hopkins's poem.

In the film *Stalker* (Tarkovsky 2002), it rains quite often and the rain does not damage the player (although there is an acid rain mod); the predominant concern in the environment of the Zone is radiation and the landscape is riddled with 'anomalies' that are caused by the Chernobyl nuclear disaster. In a remediation of the science fiction of the Strugatsky Brothers (1978) and the film, *Stalker* directed by Andrei Tarkovsky, the Zone of *Stalker: SoC* is a space fraught with danger and fear. The Zone seems to have a life of its own – in Tarkovsky, it gets an almost mystic feel but in the *SoC* also it seems to have a life of its own, later revealed to be a manifestation of an advanced AI called the C-Consciousness. Much of the concern of all the three texts is around the Zone. In the Strugatsky brothers' *A Roadside Picnic*, the Zone is defined as the debris of a picnic that an alien civilization chose to have on Earth:

"No, wait a minute." For some reason, Noonan felt cheated. "If you don't know simple things like that.... All right, the hell with reason. Obviously, it's a real quagmire. OK. But what about the Visitation? What do you think about the Visitation?" "My pleasure. Imagine a picnic." (Strugatsky and Strugatsky 1978)

The Picnic changes into a mystical Frontier in Tarkovsky where the protagonist, the Stalker, is constantly connected to James Fenimore Cooper's *Leatherstocking Tales* (1876) and is constantly called Chingachgook (a Mohican from Cooper's *Last of the Mohicans*) by one of the characters. In the game, the Zone is a possibility space. In the novel and the film, the Zone also evokes deep philosophical ruminations such as already seen in the connections of *Fallout 3's* Capital Wasteland with its literary forbears in Eliot and Hopkins. As the authors in *Roadside Picnic* say,

Or how about this hypothetical definition. Reason is a complex type of instinct that has not yet formed completely. This implies that instinctual behavior is always purposeful and natural. A million years from now our instinct will have matured and we will stop making the mistakes that are probably integral to reason. And then, if something should change in the universe, we will all become extinct – precisely because we will have forgotten how to make mistakes, that is, to try various approaches not stipulated by an inflexible program of permitted alternatives. (ibid)

Reason is seen as incompletely formed or unformed instinct in a rather brave formulation and even in the later part of the comment this incompleteness of reason is an important factor in humanity's survival. In Tarkovsky's *Stalker*, the protagonist, says

Let everything that's been planned come true. Let them believe. And let them have a laugh at their passions. Because what they call passion actually is not some emotional energy, but just the friction between their souls and the outside world. And most important, let them believe in themselves. Let them be helpless like children, because weakness is a great thing, and strength is nothing. When a man is just born, he is weak and flexible. When he dies, he is hard and insensitive. When a tree is growing, it's tender and pliant. But when it's dry and hard, it dies. Hardness and strength are death's companions. Pliancy and weakness are expressions of the freshness of being. Because what has hardened will never win. (Tarkovsky 2002)

This comment is also in the same vein as the one in the Strugatsky's novel. Pliancy can be compared to the incompleteness of reason. The Zone epitomises such a scenario of incompletely formed reason and of pliancy. In the game, too, the vast expanse of the Zone, its ambient noises and the uncertainty of what may follow is fraught with anxiety. To compare with Tarkovsky's *Stalker*, a few lines from his father Arseny Tarkovsky's poem are illustrative:

> Not a leaf was burnt up
> Not a twig ever snapped ...
> Clean as glass is the day,
> But there has to be more. (Ibid)

There has to be more. The sense of longing and waiting is important in the Zone. The Zone forms that locale where a range of potentialities become probable for the player to choose from.

In their foundational work on Game Studies, Katie Salen Tekinbaş and Eric Zimmerman (2003) described videogames as a 'space of possibility'. They state that 'we have invoked the space of possibility metaphorically, to mean an abstract decision-space or conceptual space of possible meaning. But what if we consider the space of possibility *literally* ---- as an actual 2D or 3D space in which a game takes place. In other words, one way to think of the space of possibility is as an actual narrative place' (Salen/Zimmerman 2003: 390). According to them, the chessboard is such a space as is the Go board that uses the intersection of points to determine the space of play. The 3-dimensional spaces, the concealed rooms and magical trap-doors of *Super Mario 64* are also spaces of possibility. On viewing more complicated narrative spaces in videogames, such as wasteland scenarios, the space of possibility or as I call it elsewhere, the 'zone of becoming' becomes a compelling metaphor as well as mechanic for ludo-narrative thinking.

The Wasteland as 'Any-Space-Whatever'

To further elucidate the zone of becoming, a more robust spatio-temporal exploration of this ludo-narrative space of possibility needs to be engaged in. The videogame is the locus of possibilities and choices. A compelling philosophical framework for looking at such a complex and continuously evolving scenario is, arguably, Gilles Deleuze's concept of the 'any-space-whatever' (*espace quelconque*). In his *Cinéma I*, Deleuze outlines three different parts of what he calls the movement-image in cinema. The first is perception, whereby the viewer perceives the frame of what is seen onscreen. The last is action, or the outcome that is viewed and construed by the viewer. In between is what Deleuze calls the affection-image. Alia Al-Saji describes affect as,

The delay or interruption in the body's immediate reaction allows conscious perception to arise [...] (2) The body waits before acting; it has the time to remember. In light of the delay opened up by affect, memories can beactualized and inserted into the present to help determine the future course of action .The way in which affect delays and prefigures action defines my body's hold on time – its access to memory and the openness of its future. To feel is to no longer play out the past automatically, but to imagine and remember it. (Al-Saji 2004: 221)

Continuing the thinking on affect, Sara Ahmed describes it as something that 'allows us to consider how boundary-formation, the marking out of the lines of a body, involves an affectivity which already crosses the line' (Ahmed 2013: 54). Brian Massumi describes it thus:

It is a state of suspense, potentially of disruption. It's like a temporal sink, a hole in time, as we conceive of it and narrativize it. It is not exactly passivity, because it is filled with motion, vibratory motion, resonation. And it is not yet activity, because the motion is not of the kind that can be directed (if only symbolically) toward practical ends in a world of constituted objects and aims (if only on screen). (Massumi 1995: 86)

While these scholars on affect have their own differences, there is agreement on the affect being a state between perception and action; this forms Deleuze's affection-image. Deleuze's own thinking builds upon the philosophies of Baruch Spinoza and Gilbert Simondon. Simondon also views affect 'as something corresponding to individuals, but in terms of a pre-individual potentiality exceeding the individual' (Keating 2019: 212). Deleuze refers to the vibratory motion of the not-yet activity as a motor action on a motor plate or as a throbbing. One important aspect of Deleuze's affection-image is what he calls the 'any-space-whatever'. He describes it as

Any-space-whatever is not an abstract universal, in all times, in all places. It is a perfectly singular space, which has merely lost its homogeneity, that is, the principle of its metric relations or the connection of its own parts, so that the linkages can be made in an infinite number of ways. It is a space of virtual conjunction, grasped as pure locus of the possible'. (Deleuze 1986: 109)

Ronald Bogue further adds that it is a 'virtual space, whose fragmented components might be assembled in multiple combinations, a space of yet-to-be actualised possibilities' (Bogue 2003: 80). Speaking of this space of possibilities from a Deleuzian perspective, Manuel DeLanda sees a connection with the intensive spaces that he discusses in terms of thermodynamics. DeLanda sees the space of possibility being structured by what he calls an 'attractor'. His description of virtuality is a significant entry-point to the discussion of the videogame spaces that have been addressed here. Describing the possibility space, DeLanda observes,

Of all the possible outcomes only one, or a few, become regularly actualised, a fact that suggests that the space of possible outcomes is greatly constrained, or in other words, that it has structure. While the possibilities making up this space are not real (other than in a purely psychological way) the structure of the space may be considered fully real and mind-independent. But if this reality is not actual (by definition) what is it? Deleuze's answer would be that it is virtual, not in the sense of a virtual reality (as exemplified by computer simulations, or even cinema) but in the sense of a real virtuality. (DeLanda 2002)

In the any-space-whatever, one possible outcome becomes actualised. The meaning of virtual here is that structured space where the possibilities exist. The any-space-whatever is the pure locus of the possible where linkages are made in an infinite number of ways. Deleuze views the any-space-whatever as leading directly to the time-image. In Ozu's films, spaces are 'raised to any space whatevers, whether by disconnection or vacuity' (Deleuze 2005: 15). The same can be attributed to the Zone in *Stalker* and its videogame remediation.

Although not mentioning affect, in his early work on space and time in videogames, Michael Nitsche states that

> The connection between the two [space and time] can only be understood if one includes the element of space and spatial experience in the equation. Thanks to the continuity of space the timeframe is freed to explore new configurations. This approach steps beyond the mechanical mapping of ergodic participation and game state change. Instead, space and spatial comprehension (e.g. as provided through the virtual camera) can be seen as the canvas through which the player understands time. (Nitsche 2007: 149)

Nitsche views the multiple reiterations of gameplay and, therefore, understands the space of possibilities as being closely knit with temporality and to be viewed as a space of the potential consisting of the possible, the probable and the necessary. Nitsche's diagram is much akin to the Bergsonian cone of time where the temporal states of the past, future and present coexist in a space of possibility. In a sense, the videogame space, in the many ways in which it has been understood, functions like an any-space-whatever, throbbing with possibility.

In an earlier article, I have developed the notion of the wasteland as a metaphor for videogames. Adrian Forest in his analysis of *Fallout 3* expands on this earlier analysis as below:

> I'd like to expand on the definition of Mukherjee's wastelands by saying that wasteland spaces are those spaces where the possibilities for place are expanding, where the range of possible places the space might become is increasing. And I'd contrast this with 'wilderness' spaces, which I'd describe as spaces where the range of possibilities for places the space could become are large, but narrowing or contracting, down to a reduced range of possible places the space could become. I can illustrate this distinction with two key video game examples: Fallout 3 and Red Dead Redemption. (Forest 2011)

Forest associates the Wasteland space with a degree of *increase* in possibilities and in opposition, posits what he calls Wilderness space, where the space of possibilities contracts or lessens. This article does not go into the binary of Wasteland versus Wilderness or anything similar because part of the affective nature of the wasteland as any-space-whatever lies in the visceral and emotional responses to a landscape. In non-digital narrative media, one is reminded of Michelangelo Antonioni's film, *The Red Desert* (1965), where the spiritual desolation of technology is captured in the vast industrial landscape that is almost apocalyptic in the way is portrays physical and moral desolation. What is important however, is to view how the digital game remediates the wasteland as a space of possibilities but as one where the scenario is that of disruption and what Massumi calls a 'temporal sink' where there is motion and yet no activity.

Deleuze comments on how Antonioni 'makes use of cold colours pushed to the limit of their plenitude or intensification in order to go beyond the absorbent function, which still maintained the transformed characters and situations in the space of a dream or a nightmare' (Deleuze 1986: 119). Antonioni's use of colour to create the intensified affective wasteland of *The Red Desert* can be compared to *Fallout 3*'s use of water and the *Stalker: SoC*'s use of the space of the Zone. The irradiated water in *Fallout 3* holds out the constant hope of pure water and fertility (as also in Eliot's poem) that in itself functions as an affective scene. In the *Stalker* film and its videogame remediation, the Zone functions as an any-space-whatever as has already been discussed here and in earlier research. There is also the consideration of whether these spaces, or entities within the conception of the wasteland space, are like what Deleuze describes as the time-image, which is a development on his conception of the movement-image. The time-image follows from movement but it is also something that exists as an independent (re)presentation of time. In Deleuze's description:

'Time is out of joint': it is off the hinges assigned to it by behaviour in the world, but also by movements of world. It is no longer time that depends on movement; it is aberrant movement that depends on time. The relation, sensory-motor situation ---> indirect image of time is replaced by a non-localizable relation, pure optical and sound situation ---> direct time image. (Deleuze 2005: 41)

The direct time-image is pure potentiality and it does not require an action to follow, as does the movement-image. Is such a conception of the time-image also relevant to understanding the videogame wastelands.

Framing Videogame Action and Affection

Alexander Galloway has famously commented, '[I]f photographs are images, and films are moving-images, then videogames are actions. Let this be word one for videogame theory' (Galloway 2006: 2). Galloway assumes that games when they are started involve action, both diegetic and non-diegetic. He also mentions, however, that 'when games like Shenmue are left alone, they often settle into a moment of equilibrium. Not a tape loop, or a skipped groove but a state of rest. The game is slowly walking in place, shifting from side to side and back again to the centre. It is running, playing itself perhaps. The game is in an ambient state, an ambient act' (Galloway 2006: 10). Galloway is determined to consider the ambient state as an ambient action. It can be argued, however, that the ambient state is not necessarily an act or leading to an act. In my own earlier work, I do not focus on the time-image mainly because I too focus on the videogame as an action and therefore, I concern myself mainly with the wasteland as any-space-whatever as the locale for future action. On relooking at the videogame wasteland and Galloway's own use of the ambient state, it is even more evident that the wasteland can be the site for the pure throbbing of time itself, unconnected from any concomitant action. Galloway uses the example of *Shenmue* where he sees 'the gently stirring rhythm of life'; instead of viewing this as action, however, it should be seen as affect. The wasteland in the videogame can exist as pure affect, a combination of visual and aural signs that describe time as independent of action as mentioned earlier in the context of the Deleuzian time-image. Videogames may be about actions but it can be argued that often they are more about affection as well and the gamescape makes for a zone of becoming (see Mukherjee 2015) where the process of choice-making can be observed in a state of prolonged affection.

Following Forest, however, one is tempted to ask whether the wasteland space can be applied more widely as a metaphor for videogames instead. In previous research, the wasteland metaphor has been connected mainly to first-person videogames but the applicability is arguably more ecumenical. The *Fallout* series started as the 'brainchild of some of the same developers who worked on Wasteland. Unable to secure the rights for a direct sequel, they channelled their creative vision into a new game heavily inspired by their previous work. The 1997 release inherited Wasteland's core themes: a post-nuclear world, player agency, and a struggle for survival' (Sulit 2024). Tracing a connection between the two blogger Micah Sulit describes the key characteristics of the Wasteland as follows:

Released in 1988, Wasteland offered players a unique blend of narrative and gameplay elements that were innovative for its time. The game was set in the year 2087, nearly a century after a nuclear war devastated Earth. Decades after the bombs fell, the remnants of the U.S. military formed the Desert Rangers, a group tasked with bringing order to the barren American Southwest. Players commanded a squad of these Rangers on a critical mission: venturing deep into the wasteland to recover vital supplies. You had to navigate the harsh realities of a post-nuclear world, facing off against mutated creatures, grappling with tough moral choices that had lasting consequences, and interacting with various factions all struggling to survive. What set Wasteland apart was its emphasis on choice and consequence. Players' decisions influenced the game's storyline, creating a sense of agency that was rare in RPGs at the time. Wasteland offered a dark humour and a focus on moral dilemmas that resonated with gamers, and its open-world structure, complex narrative, and challenging gameplay laid the groundwork for many post-apocalyptic games to come. (ibid)

The emphasis on choice and consequence, open-world structure and moral dilemmas in *Wasteland* were carried over in *Fallout* which added a more expanded character-building system that allowed players to forge their own path through the game. In establishing the connection, Sulit calls the connection a 'spiritual' one wherein the key mechanics of the post-apocalyptic game were developed, namely, 'a vast, explorable world, turn-based combat with an emphasis on character skills, and branching narratives driven by moral choices' (ibid). Since the popularity of the *Fallout* games, there was a resurgence of interest in *Wasteland* and this resulted in two sequels. The Real-Time Strategy element in the game adds to the space of possibility and opens up a larger set of choices. The wasteland space in *The Wasteland 2*, for example, is a vast zone of empty and desolate spaces where danger lurks in unexpected corners. As the game manual warns 'And okay, maybe we had and maybe we were – for a minute. But the Wasteland doesn't stand still, and there's never just one threat out there' (Wasteland manual 2025). In RTS games, often the ambient gamescape controlled by the AI performs actions without the player's intervention (such as the resowing of crops in *Age of Empires*) where the machine's agency is more perceptible. From the player's perspective, however, it could be seen as the ambient environment that throbs with the potential of future actions. Even otherwise, the vast possibility space of the wasteland in the RTS throbs with the slight and ambient indications of activity. Looking self-reflexively at this, *Stalker: SoC uses* the explanation of an artificially intelligent noosphere altering entity called the C-Consciousness as responsible for the creation of the Zone and its existence as a possibility space fraught with danger. Given the influence of *The Wasteland* on its core concept and design as well as its beginnings as an RTS, it is not surprising that the *Fallout* series has retained the notion of the affective any-space-whatever in the way in which it conceives videogame spaces.

The wasteland space is not constrained to a literal wasteland and as with its literary and filmic forbears, the videogame wasteland is also one that is present in the mind (as in Eliot's The *Waste Land* and Antonioni's *The Red Desert*) and by extension in other spaces that are not directly called wastelands. The battle-scarred landscape of *This War of Mine* is one such 'wasteland space'. The game is described as 'an ambitious attempt to tell human stories of civilians trapped by war, attempting to survive in combat zones' (Hamilton 2014). Players begin by playing as three civilians trying to survive in a ruined but habitable building. The gamescape is described thus: 'it is a realistic chaos. Elements out of your line of sight are obscured with a fog of war made up of delicate lines, and the only hints of colour are red orbs emanating from the sources of sound as you investigate a building, made by rats scurrying around – or people approaching' (ibid). The realistic chaos is reminiscent of the wasteland spaces in *Wasteland* and the *Fallout* games. It is within such chaos that the player is able to make choices and formulate a plan for survival.

A more recent game that comes to mind is set in Soviet Russia and presents the history of a Czechoslovak battalion trying to make their way home in a bizarrely circuitous route traversing the length of Russia from Moscow to Vladivostok via the Trans-Siberian Railway. This is a historical event that many people from outside Czechia and Slovakia may not know of but it is indeed well documented in the archives of the Museum of Czechoslovak Legions, for example. *The Last Train Home* (Ashborne Games 2023) is the story of a fictitious train that traversed the length of Russia to reach Vladivostok in a heroic struggle for survival. Much of the game is about managing a group of soldiers journeying through hostile territory fraught with imminent danger and deprivation. There is also emotional distress that makes the members of the player's group desert and sickness that causes death and physical debilitation. Resources have to be gathered by foraging in abandoned locations and sometimes nothing can be found. It is perhaps difficult to call the region a 'wasteland' but the desolation and chaos depicted in the game does present a space of possibilities much like the ones described in earlier. *Fallout 3* makes a direct connection with war: 'War, war never changes' (Bethesda 2008) as does the cinematic introduction in *Fallout 4*. The connection with war and wastelands is not new and Eliot's *The Waste Land* represents the horrors of trench warfare in the unforgettable lines 'I think we are in rats' alley/ Where the dead men lost their bones' (Eliot 2013: ll. 115-6). Paul Fussell, in his commentary on the First World War and modern memory notes that '[t]o be in the trenches was to experience an unreal, unforgettable enclosure and constraint, as well as a sense of being unoriented and lost' (Fussell 2013: 51). This sense of disorientation and loss is not present in the intensely action-packed videogame *Battlefield One* (DICE 2016) and as a commentator states 'a game set in the Great War will necessarily whitewash the horrors of trench warfare' (Hamilton 2014). There are indeed surprisingly very few games set in the Western Front of the First World War in comparison to the legion of games set in World War Two. The wasteland metaphor is key to the portrayal of the desolation and chaos of war, especially in such spaces such as the rat-infested trenches of the Western Front in the First World War and in the vast expanses through which the legion in *The Last Train Home* has to survive. Whether and how the 'rat's alley where the dead men lost their bones' or the vast desolation of Siberia can be viewed as any-space-whatevers or possibility are questions that need to be asked. Also, how the wasteland in videogames functions as any-space-whatevers and opens up a larger discussion on the way wastelands function as a narrative trope in other storytelling media is also a matter of serious consideration.

Conclusion: Re-appraising the Wasteland Metaphor in Videogames

From the above discussion, it is useful to reappraise the wasteland metaphor in videogames, extending it to other genres and titles beyond the ones that are directly about wastelands. The way in which wasteland spaces function is better understood using Deleuze's (non)philosophy of the cinema wherein the earlier formulation of understanding videogames as action can be seen to be limiting. More emphasis needs to be given to the videogame spaces as locales of affection and the space, already described as 'any-space-whatever', could be identified as the space of possibilities that earlier Games Studies research has posited. Of course, it could be tempting to construct a binarism between wasteland games that function as any-space-whatevers and those that are more about instant action but this article does not wish to engage in such a definition. Rather, the aim here is to study the wasteland metaphor as a way to 'stretch' out the gameplay of videogames and explore the medium's workings as a space of possibilities; it can be argued that even in a game that presupposes fast-paced action the player makes choices within a space of possibilities in what may be described as an affective scenario, in between perception and action. The wasteland space makes such a phenomenon more obvious. The affective nature of the game arguably provokes more thinking about fragmentariness, disorientation and the dystopic while also creating a scenario where a multiplicity of action can be contemplated and planned. Also, in themselves, the wasteland scenarios are sites of throbbing potentiality that can often continue to be so if the game algorithm or the player does not act. In this sense, the wasteland may serve as an illustration of the Deleuzian time-image which can be interpreted as pure potentiality. Of course, in most gameplay experiences, the any-space-whatever is the locus of the affection-image and thereby, of action. As shown earlier, the wasteland metaphor itself may have certain constituent elements, such as water in *Fallout 3* and the Zone in *Stalker: SoC*, that are crucial in the creation of the wasteland (literally, as the wasteland in *Fallout 3* is devoid of pure water and that in *Stalker: SoC* is what it is because of the Zone created by the AI called C-Consciousness.

While this article does not aim to pursue a typology of the wasteland, it nevertheless reemphasizes the importance of the wasteland as a metaphor in understanding the possibility space in videogames. Unlike the OTT version of *Fallout*, where the wasteland seems to be more the site of action rather than affection, the videogames in the series, arguably, present a much more affective scenario that presents a multiplicity of possibilities and perhaps remediate the idea of the wasteland as an affective locus that is present in earlier narrative media such as Eliot's poem or Antonioni's film. When watching *Fallout on* television, although the scripted events seem limiting, nevertheless the wasteland environment opens up a possibility space where a multiplicity of future actions can be imagined and where the story viewed onscreen is but one of many possible *Fallout* stories. In the videogame, of course, the player's story is just one of the many that are possible and the wasteland space works both as a mechanic and a metaphor to enable the space of possibilities that influences the choice-making and actions that make the videogame a more deliberately writerly medium.

References

Bianchi, R. (2024): 'Beyond Barren Wastelands: The Greening of the Post-apocalypse in Video Games', *Journal of Games Criticism*, Volume 6 Bonus Issue A.
Bogue, R. (2003): *Deleuze on Cinema*. New York: Routledge.
Bolter, J.D./Grusin, R. (2000): *Remediation: Understanding New Media*. Reprint edition. Cambridge, Mass.: The MIT Press.
Bradbury, R. (1989): *There Will Come Soft Rains*. Perfection Learning.
Cooper, J.F. (1876): *The Leather Stocking Tales*. Boston, MA: Houghton, Mifflin and Company.
DeLanda, M. (2002): *Intensive Science and Virtual Philosophy*. London, New York: Continuum.

Deleuze, G. (1986): *Cinema 1 : The Movement-image*. Translated by H. Tomlinson and B. Habberjam. London: Athlone.

Deleuze, G. (2005): *Cinema 2: The Time Image*. London: Continuum International Publishing Group Ltd.

Eliot, T.S. (2013): *The Waste Land*. Martino Fine Books.

Forest, A. (2011): 'The Waste And The Wild', *Three Parts Theory*, 4 April. Available at: https://threepartstheory.wordpress.com/2011/04/05/the-waste-and-the-wild/ (Accessed: 16 January 2025).

Fussell, P. (2013): *Great War and Modern Memory: Perspectives from Philosophy, Theology, and Psychology*. New edition. Oxford: OUP USA.

Galloway, A.R. (2006): *Gaming: Essays On Algorithmic Culture*. Minneapolis: Univ Of Minnesota Press.

Hamilton, M. (2014): 'This War of Mine – gaming's sombre antidote to Call of Duty', *The Guardian*, 10 October. Available at: https://www.theguardian.com/technology/2014/oct/10/this-war-of-mine-gamings-sombre-antidote-to-call-of-duty (Accessed: 16 January 2025).

Il Deserto Rosso (1965): Film Duemila, Federiz, Francoriz Production.

Keating, T.P. (2019): 'Pre-individual affects: Gilbert Simondon and the individuation of relation', *Cultural Geographies*, 26(2), pp. 211–226.

Massumi, B. (1995): 'The Autonomy of Affect', *Cultural Critique*, (31), pp. 83–109. Available at: https://doi.org/10.2307/1354446.

Mukherjee, S. (2010): '"The Water of Life Freely": Water and the Wasteland in Fallout 3', *Ludus Ex Machina*, 31 January. Available at: http://readinggamesandplayingbooks.blogspot.com/2010/01/water-of-life-freely-water-and.html (Accessed: 13 January 2025).

Mukherjee, S. (2015): *Video Games and Storytelling: Reading Games and Playing Books*. Basingstoke: Palgrave Macmillan.

Nitsche, M. (2007): 'Mapping Time in Video Games', in *DIGRA*. Tokyo. Available at: www.lcc.gatech.edu/ nitsche/download/Nitsche_DiGRA_07.pdf.

Salen, K./Zimmerman, E. (2003): *Rules of Play: Game Design Fundamentals*. Cambridge, Mass: The MIT Press.

Stalker [1979] (2002). Artificial Eye.

Strugatsky, A./Strugatsky, B. (1978): *Roadside Picnic, Tale of the Troika*. Translated by A.W. Bouis. New York, N.Y.: Pocket Books.

Teasdale, S. (2012): *The Collected Poems of Sara Teasdale*. Digireads.com.

Wallace, I. (2009): 'Why doesn't it ever rain? – Fallout 3', *GameFaqs*. Available at: https://gamefaqs.gamespot.com/boards/939933-fallout-3/48207622 (Accessed: 15 January 2025).

Wasteland manual (2025): *Wasteland Wiki*. Available at: https://wasteland.fandom.com/wiki/Wasteland_manual (Accessed: 16 January 2025).

Digital Games

Ashborne Games (2023) 'The Last Train Home'. Vienna: THQ Nordic. Available at: http://www.amazon.co.uk/dp/B0017Y380C.

Bethesda Softworks (2008) 'Fallout 3'. Bethesda. Available at: http://www.amazon.co.uk/dp/B0017Y380C.

DICE (2016) 'Battlefield One'. Vienna: Electronic Arts. Available at: http://www.amazon.co.uk/dp/B0017Y380C.

Gameworld, G.S.C. (2007) 'S.T.A.L.K.E.R. : Shadow of Chernobyl [DISC]'. California: THQ. Available at: http://extralives.wordpress.com/game-reviews-shooters/stalker-shadow-of-chernobyl/.

Gaming in Ruins
Alienation and Hope in *NieR: Automata*

Timothy Welsh

Abstract

PlatinumGames' NieR: Automata has received significant critical attention for its exploration of posthuman existentialism. This article builds on that work by situating the game's meditation on alienation within a broader industry shift away from single-player, narrative-driven games toward highly monetized live-service models. Drawing on recent scholarship on video game ruins, it links the game's depiction of civilization's collapse to gaming's own entanglement with the end of capitalism. In this context, what Automata *asks players to do takes on greater significance – not just within its fiction, but within extractive economies structured by alienation. Through role-playing mechanics that invoke neoliberal imperatives,* Automata *implicates players in the objectivized logics of contemporary digital culture. And yet, it also offers small but significant opportunities to resist those logics. The article concludes with a focus on Ending E, which models a prosocial, posthuman community and a hope that the future might unfold differently.*

Platinum Games's *NieR:Automata* (2017, hereafter *N:A*) begins with human civilization already in ruins. Buildings are destroyed, roadways are crumbling, abandoned cars are overgrown with vegetation, and there are no humans to be found. The player is told that, when aliens invaded Earth, the surviving humans fled to the Moon, while the human-like androids – the YoRHa – remained to fight the alien's machine army "for the glory of mankind." It is later revealed, however, that humanity's escape to the Moon is a lie and both the humans and aliens are long extinct. Even so, the android/machine war rages on, each side executing their programmed directives in perpetuity. The surviving androids, like the playable 2B and 9S, are relics of a conflict that is simultaneously over and ongoing, persisting among and as the ruins of the human civilization they were created to protect. The game's dramatic conflict is propelled by 9S as he confirms humanity's extinction and reckons with the futility and pointlessness of continued struggle. Before it became a formally complex, existential reflection on surviving in the ruins, though, *N:A* almost became a very different kind of game.

The previous game in the *Drakengard/NieR* franchise, Cavia's *NieR* (2010), was released at the tail end of what Felan Parker (2018) identifies as a period of heightened discourse around games as art. Despite poor sales, the franchise retained enough of a cult following to warrant consideration for a sequel. Two proposals emerged: one for the narrative-driven, action-adventure game that became *N:A*, and another for a mobile farming simulator on the PlayStation Vita, designed in the style of Zynga's *FarmVille* (2009; cf. Robinson 2017). This potential future of *N:A* is notable because of the influence *FarmVille* had on an industry in flux amid what Jesper Juul has called the "casual revolution" (2012). The success of "casual" games like *FarmVille*, even during the Great Recession, knocked the industry on its "collective ass" and led to the broad adoption of similar design and monetization strategies across genres (Schell 2010). knocked the industry on its ass and led to the broad adoption of similar design and monetization strategies across genres. The same year *N:A* was released, EA closed Visceral Studios, a move widely interpreted as a "death knell" for single-player, narrative-driven games (Sarkar 2017; Gamasutra Staff 2017). EA's justification for this decision cited a "pivot" toward live-service models, reflecting an industry-wide shift toward recurrent user spending, microtransactions, and perpetual engagement loops.

Given these market conditions, it is a wonder that *N:A* emerged in its current form as a single-player, story-focused, philosophical experience without microtransactions. Its alternative possible future as a *FarmVille* clone, though, places it as an inflection point in the history of gaming. Through the embrace of live-service models, the industry expanded and further monetized "the competition, repetition, and quantified objectives that make up gamified designs," which, Patrick Jagoda (2020) argues, "correspond with some of the most pernicious aspects of advanced capitalism" (12). Situating *N:A* within this broader industry shift joins its depiction of civilization's collapse to gaming's own entanglement with capitalist ruins. In this context, what the game asks players to do, particularly in moments like Ending E, take on greater significance—not just within its fiction, but in the context of extractive economies structured by alienation.

Anna Tsing (2017) offers a useful explanation of the role of alienation in the "history of the human concentration of wealth" leading to capitalist ruins (5). For Tsing, "making both humans and nonhumans into resources for investment" is achieved by imbuing them with "alienation," or an "ability to stand alone," which turns them into "mobile assets." Alienation restratifies the surrounding environment such that "only one stand-alone asset matters: everything else becomes weeds or waste" (ibid: 6). Once the desired asset is mined out, what remains can be abandoned: "thus, simplification for alienation produces ruins, spaces of abandonment for asset production." Tsing's study focuses on deforestation and the mushroom industry; however, video gaming perpetuates similar forms of alienation for investment across several domains. Materially, gaming relies on the extraction of natural resources like coltan for hardware production as well as energy for live-service connectivity, contributing to environmental degradation (cf. Lasker 2008; Joseph-Gabriel 2015; Mills, et.al. 2019). The industry is notorious for exploitative, dehumanizing labor practices, including crunch culture and outsourcing (cf. Dyer-Witheford/DePeuter 2006; Bulut 2020). Art and story assets are set up to be infinitely extensible through transmedial franchising and cross-collaborations as well as the packaging of costumes, characters, maps, and entire episodes into discrete, sellable downloadable content (DLC) packages (Kline, et. al. 2003; Jenkins 2006). Formed into neoliberal subjects through gamified designs, players are also "mined" for attention (cf. Ash 2012; Möring/Leino 2016; Jagoda 2020; Odell 2020; Soderman 2021; Hayes 2025). Following the casual revolution, design strategies emphasize what Schell (2010) calls "mental tricks" like the "fear of missing out" (FOMO) and sunk-cost fallacy to foster compulsive engagement and rationalize microtransaction purchases. The result is often grindy, repetitive gameplay that can feel like work (cf. Fuchs 2014; Sun 2022). In response to these forms of "playbour," players often seek out optimized strategies to minimize the grind and get the most value for the investment of time or money (Kücklich 2005; cf. Paul 2024). As play collapses into, as Ergin Bulut (2013) puts it, "nothing but dead labor," it is no wonder these trends coincide with widespread player sentiment that live-service games "don't respect their time," itself an ironic formulation intertwining play and productivity (410; Dookeran 2024).

To play a video game is gaming in these ruins, subject to and participating in modes of alienation that characterize the end of capitalism. Games set in post-apocalyptic scenarios, like *N:A*, hold special relevance in this context because they depict ruins, which Walter Benjamin (2002) famously describes as a form of "dialectical image" (473). Decayed figures of the past and decaying in the present, he suggests, have the potential to disrupt the capitalist myth of progress. Several scholars have applied Benjamin's framework to the post-apocalyptic ruins of video games. Emma Fraser (2016), for example, argues that video games offer up ruins as playable landscapes and, thus, have a unique capacity to subvert teleological histories. Fraser and others recognize, though, that video games tend to position players as the saviors of fallen worlds, a trope that can overlay a conservative teleology on virtual ruins. Óliver Pérez-Latorre (2019) observes the presence of this trope in several of the best-selling post-apocalyptic video games released in the wake of the Great Recession, the period coinciding with the industry's infamous pivot. The concern with the prominence of the trope, he explains, is that video games and other forms of popular fiction "influence the construction of the social imaginary of the recession around concepts like austerity, precariousness and models of heroism depicted as virtuous for dealing with scarcity and as a way out of global crises." Tsing (2017) too takes issue with "popular American fantasy" foregrounding an understanding of survivalism that involves "saving oneself by fighting off others" (27). In its place, Tsing urges a shift to strategies of "collective survival" that emerge from, not teleological narrative, but "a rush of troubled stories" about our "contaminated diversity" (34).

The *Drakengard/NieR* post-apocalyptic transmedia universe, featuring several games with multiple-ending structures and numerous collaborations, could certainly be considered a "rush." As such, though, it can be quite unwieldy, making it difficult to address comprehensively or distill a unifying trope, utopian or otherwise.[1] This inconsistency in the franchise extends to the critical literature as well. Though Caighlan Smith (2024) observes in *N:A* a tendency to recenter the human, Alexandre Paquet (2021) has shown that *N:A*'s deployment of a complex, multiple-ending structure resists reinscribing a teleological history and instead suggests "that hope and utopia are intimately tied to one's willingness to engage in a broader collective" (127). Critical assessment of *N:A* is further complicated by its recent legacy of participating in uneven collaborations with a wide variety of other game franchises, including gacha-style mobile games that sell sexualized *N:A* art assets as microtransaction content. The modest goal of this essay, then, is to track one ending, Ending E, and frame it in relation to the game's transmedial expansion and the neoliberal strategies associated with commodified play. Even if the broader franchise cannot maintain it indefinitely, Ending E offers a fleeting but significant opportunity to reject the objectified logics of self-contained, alienated individualism by raising the possibility of collective survival instead.

Though there has been a recent wellspring of critical interest in *N:A*, these conversations have tended to focus on its engagement with themes of existentialism and posthumanism (cf. Gerrish 2018; Jacevic 2017; Wahba 2018; Lima/Walesa 2022). Such discussions are crucial to understanding *N:A*'s contribution to the social imaginary Pérez-Latorre describes; however, they often do not take up and potentially abstract the game's relationship to capitalist ruins or the material conditions of gaming itself. Even as *N:A* invites players to make choices that hope for an alternative to alienation, it is necessary to acknowledge that *N:A* remains a commercial product: conceived as a *FarmVille* game and part of a sprawling transmedial franchise, which includes licensed mobile assets that participate in predatory monetization practices. *N:A*, then, does not merely represent the end of capitalism as a setting for posthumanist reflection on alienation but is itself imbricated in the ruination. Rather than compromise the entire project, however, *N:A*'s position within the ruins of gaming recommends it for what Philip Wegner (2020), drawing on Fredric Jameson, calls "creative reading," or, perhaps more appropriately for a game, creative play (15). To that end, this essay situates the game's representation of posthuman survival among the ruins of human civilization in the material context of the ruins of capitalism from which it emerges and

---

1   I am indebted to YouTubers like Clemps (2015-2018) and anonymous contributors to wiki pages (cf. "Nier Wiki" 2024; "Arc" 2024) for my admittedly limited understanding of the *Drakengard/NieR* lore. Where relevant I have included extended summaries, both to contextualize the discussion as well as to compile it in one place to aid future researchers.

with which it interacts. Through the implementation of role-playing mechanics that align with neoliberalism, *N:A* actively implicates players in the alienating human-machine assemblages characteristic of contemporary digital culture. Even so, throughout the game – and especially in Ending E – players are given opportunities to reject the objectified, dehumanizing logics of late capitalism. These gestures may appear minor. Playing *N:A* "creatively," though, means attempting to hold on to precarious moments like these, which model a prosocial, posthuman community and, in Wegner's sense, evoke hope that the future might unfold differently.

Despite Ruination

While it is clear that *N:A*'s Earth has lain abandoned for a long time, it is equally clear that humanity is never coming back. Robert Yeats (2021) observes, "Rather than a creative destruction, a concept derived from Karl Marx which proposes that new capitalist economic orders arise from the ruins of the old, the urban destructions of post-apocalyptic fiction frequently appear total and lasting, beyond repair or reclamation" (1). It certainly feels like that is the case here as well. Though still recognizable as (unspecific) urban ruins, the game's landscapes of ruination are cut off from their prior roles in the social lives of humans. The industrial port and abandoned factory–the first settings the player encounters – are rusted over. Piles of discarded iron and broken components lay where they fell. Much of the machinery remains operational, appropriated and maintained by the machines for the fight against the YoRHa androids. For example, portions of the port attached to the factory have been reassembled into a miniboss, which forms the arm of the massive area boss, respectively named Marx and Engels. Not far away, the androids stake out an enclave in the city ruins. Their tactical hideout is situated between derelict buildings, the sturdy remains of the city grid. In these urban ruins, massive mounds of concrete rubble from buildings and overpasses have begun to cover over with grass and flowers, and moose roam freely through what used to be streets. Christian Hines (2021) points out that 9S and 2B have noticed Earth is "rewilding." Even the remains of the machine war have become "untimely ruins" reclaimed by the Earth (Yablon 2010: 12). Out near the ancient ruins of the Forest Castle can be found massive, incapacitated machines, slouched over and covered in flora. We are fully in the post-anthropocene, and humanity's technological progress did not secure its survival.

Perhaps the best example of this failure of progress is the surviving androids, themselves a ruin of humanity. In an effort to survive the global spread of White Chlorination Syndrome (WCS) in the early 2000s – an incurable disease that slowly turns the cells of the afflicted to sodium chloride – Project Gestalt developed a process to separate a person's soul, referred to as a "Gestalt," from their body and preserve it while WCS ran its course. Once the disease had passed, the Gestalt could be reunited with a new body, called a "Replicant," grown from the original genetic material. Androids were created to oversee this process, stewarding the Replicants until they could be inhabited by their Gestalts. The project failed, however. Over the thousands of years it took to eliminate WCS, the Replicants developed their own personalities and cultures. When the Gestalt awakened, most were rejected by their associated Replicants. The rejected Gestalts lost sentience and became hostile. Their "relapse," as it was called, triggered the corruption of their associated Replicants – a condition known as "Black Scrawl" due to strange bands of black text that spread across the victim's body, indicating their degradation. The Gestalts went extinct in 4198. Afterward, the androids gathered the genetic data from Project Gestalt and stored it on the Moon, hoping to find a way to revive humanity. In the meantime, they remained locked in an endless battle with the alien machine army. By the time *N:A* begins, the androids – once the pinnacle of human technological achievement, unable to resuscitate their creators—have been reduced to caretakers of the genetic remnants of an extinct species.

It is worth noting the resemblance of the origin of the apocalypse in the *Drakengard/NieR* franchise to the bombing of Hiroshima and Nagasaki. WCS entered this universe during the events of *Drakengard* (2003) when The Giant and The Dragon transitioned between dimensions to appear in Tokyo in 2003. After the Dragon defeated The Giant and was subsequently shot down by the Japanese airforce, their bodies disintegrated into a white substance that covered the city like snowfall. This substance was called "maso" and would ultimately be found to be the cause of WCS. Originally contained through quarantining infected citizens behind a wall, WCS spread globally after a series of nuclear strikes across Japan intended to eradicate the disease instead distributed maso into the atmosphere. The bombings of Hiroshima and Nagasaki loom large in the cultural imagination of Japan for obvious reasons (cf. Tanaka 2014). Tsing (2017) also locates at that historical moment the recognition of a perpetual state of precarity, both existentially and economically. Technological advancement had not made us safer; instead, we "could destroy the livability of the planet – whether intentionally or not," and global capitalism had not brought universal financial stability, for "everyone depends on capitalism but almost no one has what we used to call a 'regular job'" (Tsing 2017: 3). From that point to the present moment, she argues, "we are stuck with the problem of living despite economic and ecological ruination" (ibid: 19).

This question of how to go on living in the ruins is central to *N:A* specifically. For instance, each of the machine factions the player encounters approach this question differently. Having separated from the machine network, these factions have to come up with their own guiding principles. They do so by imitating aspects of humanity they gleaned from studying their history, and literally taking up residence in their ruins. The machines in the ancient ruins of the Forest Castle created a monarchical civilization, installing the memories of the deceased Forest King in the infant robot, Immanuel. Another faction apes religion, forming a death cult in the abandoned laboratory. The machine imitations are tragically abstracted from the human history that generated them, far too simplistic to survive. Even poor Pascal's pacifist commune illustrates what Jenny Odell (2020) calls the "impossibility of retreat," eventually getting caught up in the fighting anyway, retreating back into the abandoned factory, and succumbing to fear (30). In coming up short, though, each faction prompts the player to consider the logical extensions of grand narratives amid physical testaments of their failure to stave off precarity.

In this way, *N:A*'s post-apocalyptic ruins do not carry on the work of the dialectical image merely through representation of the end of capitalism. They become occasions to both confront abstracted history, as imitated by the machine factions, as well as consider other forms of communities of survival. Though the machine imitations are tragically incomplete, they subvert too-easy solutions that might shortcut the work of the dialectical image of the ruins. As itself an artifact of the ruins of gaming, *N:A* turns this consideration back on the player as well, using its mechanics to prompt introspection about the logics of the task of gaming. For this to work, though, it first has to convince players to inhabit those logics.

Neoliberal NieR

At the level of gameplay, *N:A* presents itself as a typical major-studio action-adventure with core elements from the role-playing game genre. As Pérez-Latorre and Oliva (2019) argue in their analysis of *BioShock Infinite* (2013), such games often reflect neoliberal gameplay structures. They describe a fundamental alignment between neoliberalism and what they call "certain 'traditional' or 'habitual' game design forms" (796). Using a health potion, for instance, is a form of quantified self-improvement through consumption of goods. *N:A* employs many of these design elements, reproducing in the player the alienation of modern digital life.

Once the tutorial mission is complete, *N:A* invites players to explore at their own pace an open world of apparently free movement and self-determination. Even once missions are triggered, the player can abandon them by simply walking away. Doing so in certain missions can trigger alternative endings, indicating the game's responsiveness to the player's self-determined paths. To meet the game's many challenges, *N:A* implies that players should strive toward self-optimization. Through the plug-in system players customize their abilities and performance. This performance is quantified in a variety of ways. Players earn experience points to level up and raise the playable character's numerical attributes. The game records the player's achievements such as number of enemies killed, number of weapons collected, and the percentage of in-game missions completed. Weapons and plugins have their own numerical levels and statistics. Attacking an enemy produces a cascade of damage numbers, allowing players to evaluate the performance of their current weapon and plugin loadout.

Our playable characters have limited plugin capacity. The stronger the plugin the more capacity it takes up, which means players have to decide not only which plugins to use but also balance plugin strength against its size. Players can even uninstall plugins relating to elements of the heads-up display (HUD), such as the minimap, damage numbers, enemy or personal health bars and so on. This allows them to both customize the kinds of information feedback they receive in game as well as balance the value of this information against the plugin capacity it takes up. For instance, they may decide they can sacrifice seeing the enemy health bar in order to squeeze out a little more capacity to fit a bigger damage plugin.

The plugins themselves can be improved and optimized. Each plugin has a numerical rating corresponding to its quality. Plugins of the same type can be merged at a vendor, raising their rating number, their statistical improvement, and their size. Plugins of the same rating do not all require the same capacity space, however, so players will need to acquire plugins with a high rating but low capacity draw. Plugins with the optimally low size for their rating are marked with an asterisk. Merging these plugins with others of similar optimal size will raise their rating. For the best performance, the player needs to acquire, arrange, and upgrade plugins, carefully merging them to improve their performance while maintaining a low capacity draw so they can fit as many as possible into the available capacity.

The imperative toward self-optimization is conducted through consumption and participation in the gameworld economy, even in this post-apocalyptic setting. The player's limited plugin capacity can only be enlarged by purchasing upgrades. In addition to the passive buffs of plugins, there are also active, temporary buffs from consumables. Though the player may arrange their plugins at any time, the plugins themselves can only be merged at a vendor. Weapons too can be upgraded only at blacksmiths in exchange for specific resources and currency. Some weapons can be found through exploration, while others are for purchase only. Certain in-game content, specifically most of the Emil storyline that goes a long way to filling out the relationship between *N:A* and the rest of the *Drakengard/NieR* universe, becomes available only after players acquire and fully upgrade every weapon. Thus, even in a post-apocalyptic context in which humans have become extinct, consumption becomes not only a way to improve performance, upgrade weapons, and customize the playable character, but also necessary to make progress and understand the story.

On the level of mechanics alone, then, *N:A* participates in the same structures of and inducements toward quantified self-fashioning through consumption that define the action-adventure role-playing game genre. What sets *N:A* apart is its self-reflexive story, which folds back on these gameplay practices and the logics of objectivization they invite players to inhabit. In other words, these kinds of mechanics typically hail players into gameplay styles similar to Hocking's (2007) assumption about the original *Bioshock* (2007): "seek power and you will progress." Following that implied gameplay style will lead the player of *N:A* to the game's final battle. Unlike *Bioshock*, which can't imagine an alternative, though, *N:A* will prompt the player to reject those logics and choose other values or forgo the conclusion (cf. Wysocki/Schandler 2013). In this way, it constitutes a kind of critical ludonarrative dissonance, reproducing contemporary gaming in order to ultimately offer an alternative (cf. Murphy 2016).

Contextualizing Ending E

The narrative structure of *N:A* is itself alienating and posthuman. It is fragmented, opaque, self-referential, and difficult. Much of the narrative context is developed through the sprawling *Drakengard/NieR* franchise, which includes several other games, DLC, art books, short stories, interviews with Yoko, and stage plays. Like previous games in the series, *N:A* requires numerous playthroughs and shifts in focalization to access multiple endings and complete the game's story arch. There are 26 endings in *N:A*, one corresponding to each letter of the alphabet. As the player unlocks each ending, their save file is annotated with the corresponding letter. Another in-game collectible, potentially subject to an objectivized logic of trackable, quantifiable progress. The only way to collect them all, though, is to allow the game to delete the file and with it the record and digital byproduct of the player's significant immaterial labor.

That the game puts players in a position in which they would willingly give up this valuable investment of time and the agency it affords by enabling replayability, is the game's greatest achievement. It brings players to this choice through subtle shifts in the ludic contract as 9S and the player learn more about what is going on. While one might read the game as setting up the kind of ludonarrative dissonance that inspired Hocking to coin the term, *N:A*, unlike the Bioshock games, ends with an actual choice, and this choice has real, material consequences for the player and other gamers. Thus, where Pérez-Latorre and Oliva fault the Bioshock series for failing to imagine anything different, *N:A*'s Ending E affirms that–as the mad machines chant–this cannot continue with a prosocial gesture toward other possibilities.

As the game begins the narrative and ludic contract appear to be aligned in a seemingly uncomplicated opposition between the human's androids and alien's machines. The player operates YoRHa androids fighting on behalf of humans against the machine army mobilized by the invading aliens. They fulfill this role by deploying combat mechanics, improving performance through the leveling and plugin systems, all of which are consistently incorporated into the narrative schema. As the player learns more about the world and the history of the battle, though, that assumed orientation is disrupted.

The first revelation comes during the initial playthrough that results in Ending A. 9S and 2B discover the corpses of the alien invaders deep underground. They are told by the highly-evolved machines Adam and Eve that the machines revolted and killed their alien creators long ago. While this information alters the scenario, the organization of the opposition is unchanged. The YoRHa and the player must still defend humanity against the alien machines. The more significant disruption occurs during the Ending B playthrough. 9S finds records that indicate that the Council of Humanity, the supposed governing body of the human colony on the Moon, was created as a part of Project YoRHa, not the other way around. When he confronts the Commander, she admits that "mankind no longer exists."

None of the remaining androids outside of the Bunker know this fact. As news of humanity's extinction leaked out, androids who learned the truth became despondent, as they no longer had a reason to live or fight the machines. Project YorHa, of which 9S and 2B are a part, is, in fact, an elaborate scheme to convince the remaining androids that humanity has escaped to the Moon. The Council of Humanity is just a server broadcasting pre-drafted directives and encouragement to give the androids a reason to go on.

Up until this point, the player's orientation and purpose within the game had been clear. The knowledge that the humans, like the aliens, are gone eliminates the narrative motivation for fighting the machines. Project YoRHa gave the android troops, and by extension the player directing them, a reason to continue the war. Once 9S and the player find out the truth of the scheme, that motivation dissolves, leaving both in search of an orienting purpose for the remainder of the game.

There is not much time to contemplate these circumstances, though, as shortly after learning the truth from the Commander 9S is deployed for the boss fight that results in endings A and B. At the beginning of the next playthrough all YoRHa androids connected to the server contract a logic virus and perish. Eventually it is revealed that one of the founders of Project YoRHa -Number 9, a precursor to 9S – installed a backdoor to the YoRHa network that would open to the machines at an appointed time. This would allow the machines to wipe out all members of the Project, destroy the Bunker, and trigger the Pods that accompany the agents to delete all information and records from the Project. No.9 believed the complete eradication of the Project and all records associated with it would ensure no androids ever learned the truth about the fake Council of Humanity.

In other words, not only does the player participate in the protection and distribution of this lie, their efforts are designed to fail, subverted from the very start. 9S is especially victimized by this betrayal. As an advanced scanner module, his job is to seek out useful information about the machines that can aid in the war effort. But he cannot be allowed to find out the truth of the lie. To prevent him from getting too close, he is accompanied by 2B. Though 2B clearly cares for 9S, it is revealed late in the game that her actual model number is 2E, which designates her role as executioner. By the time events of *N:A* begin 2B/E has killed 9S 47 times. Each time, 9S's consciousness is reloaded from the Bunker server and he starts his reconnaissance – and relationship with 2B/E – all over again.

It turns out the machines are in on this scheme as well, sustaining the war for their own self-evolution. The machines have come close to winning the war previously, however, they interpret the prime directive of their programming – defeat the enemy – as requiring an enemy to fight. Thus, rather than eliminate the androids outright they kept the war going, even delaying use of the backdoor, so the escalating conflict with the YoRHa would continue to apply evolutionary pressure. The YoRHa use combat data to keep improving their models and tactics against the machines, which makes them a worthy adversary, who keeps improving to match the self-adaptations of the machine network. The android and machines – both of their creators extinct – sustain a cycle of perpetual war, each side enhancing itself, becoming more efficient, solely to maintain conflict for the sake of itself.

The full revelation of the significance of the player's role in this conflict recontextualizes the game's invitation to self-fashioning toward optimization. The entire enterprise is futile, doomed from the start and restart, in service, not of progress toward victory, but unending war. The player's efforts perpetuate the inspiring lie of humanity's survival by fulfilling the computational logic of the machines who need an enemy to keep improving in order to force their own evolution.

Here, in the breakdown of the myth of progress, *N:A* opens an opportunity for the player to embrace an alternative to this endless struggle. Some of the machines the player met along the way have suggested the possibility of other modes of being. The disconnected machines like Pascal, the Forest King, and Adam and Eve are anomalies created by the machine network as a result of intentional self-degradation to test other possibilities. These "diversified vectors of evolution" are all wiped out over the course of gameplay as the network tries to correct itself. Even so, they demonstrate that the machines are looking for a way out of this cycle as well. In fact, at the completion of Ending D, the machines decide to leave Earth to find their own home and invite 9S to join them. Continuing on to Ending E, though, the player is given an opportunity to break the cycle without retreating from the planet by reuniting 9S and 2B.

At the completion of Project YoRHa, the Pods must carry out the deletion of all data pertaining to the project. At this point Pod 042 refuses and with the player's permission – phrased as a question from Pod143, "do you want them to survive" – the two pods commence data recovery. In order to succeed, the player must survive an extremely difficult bullet-hell minigame, a genre *N:A* invokes throughout to represent hacking, in which they must shoot the scrolling credits. The game's most challenging segment, then, involves firing back at its creators to prevent the ending. Yet, as the player shoots down their names, the voices of those who worked on the game can be heard on the non-diegetic soundtrack joining the chorus of the anthemic "Weight of the World" as if cheering the player on.

Upon dying in this sequence, the player is prompted with questions to which the player must respond NO in order to continue:

DO YOU BELIEVE GAMES ARE SILLY LITTLE THINGS?
DO YOU BELIEVE THIS WORLD HAS NO MEANING?
GIVE UP HERE?

As the player decides on an answer, messages from other players who have completed the game scroll across the screen. The player can then accept a rescue offer from other players, which adds more ships and more firepower. When these extra ships are killed, a message appears stating that the rescuer's data has been lost, the meaning of which is unclear at that point. With the added help, the bullet-hell sequence becomes (more easily) winnable and players move on to the concluding cutscene.

Having retained the data and thus memories and identity of the playable androids, the Pods reconstruct them. Pod 153 asks Pod 042 if rebuilding them from parts of the same design would not "simply lead to the same conclusion as before." Pod 042 replies that he cannot deny that it is possible but that "the possibility of a different future also exists." "A future is not given to you," he posits, "It is something you must take for yourself." Immediately after Pod042's hopeful response, which uses the second person to affirm individual agency in systemic change, Pod153 invites the player to send a message to other players "who are suffering because they cannot finish NieR: Automata." These messages – composed from a collection of prewritten phrases – are the ones that appear after one fails the bullet-hell credit sequence. Then, Pod042, noting that the player died multiple times attempting the ending, observes "You faced crushing hardship, and suffered greatly for it" and asks "Do you have an interest in helping the weak?" By selecting "YES," the player agrees to come to the aid of another struggling player, "however, in exchange, you will lose all of your save data." After pointing out this condition, Pod 042 reiterates the question: "Do you wish to rescue someone – a total stranger – in spite of this?" As Pod 042 continues to confirm the player's intention, the player can of course select "NO" at any point, return to the title screen, and save their game. If they respond "YES" to all of Pod 042's questions, they are shown their data as it is deleted and they are added to the roster of rescuers who show up randomly to help someone struggling with the final sequence.

The game's ludic and narrative contracts come back into alignment at this moment. As Caymus Ducharme (2021) argues, "Ending E offered a solution to mend the story-gameplay divide" by "[allowing] the player to apply the theme the ending taught them and generate a real effect on real people. Through interactivity, Ending E does more than say something in a more effective way: it convinces people to act, to go beyond the game and leave an impact on the real world." Though one can complete the sequence solo, the crushing hardship of the game's final challenge becomes an opportunity for the player to enact Pod042's possibility of a different future. It is an exchange of the repetitive, futile labor that produced the file for the opportunity to connect with another player, even in such a small way. With this choice, players release the by-product of perpetual self-improvement in response to and driving non-stop competition, give up control of 9S and 2B and the option to reload a save file in which they are still playing out their cyclical role in the machine war. Deleting one's save file leaves them to pursue their own self definition and meaning and enables another player to have the chance of making the same choice.

It is not much, just helping a stranger beat a video game. Against the ills of gaming culture and its participation in an all encompassing attention economy, such small moments can easily be swallowed up. Fraser (2016) points out, after extolling the critical potential of playing among the virtual ruins in video games, these are still commercial products (190).[2] Indeed the legacy of *N:A* has tended more toward the sexualization of 2B, an emotionally complex character reduced to a mobile asset. Even so, *N:A*'s participation in the very ruins of gaming makes this inflection point that is Ending E all the more significant for how rare and fragile it is.

As *N:A* grows older, and fewer players take it up and even fewer find Ending E, there will be fewer new messages added to the database. All that will remain of this inflection point will be digital traces of players who already left the world of *N:A* behind (cf. Janik 2019). These messages from a remote server meant to inspire hope and perseverance, confront the player like the fabricated communications of the Council of Humanity. Still, prevailing against the game's ludic and narrative challenges, players are rewarded with a choice to rescue someone rather than preserve data. Deciding to exchange the repetitive, futile labor that produced the file creates an opportunity to connect with another player and, even in such a small way, to relieve digital alienation.

Creative Collaborations

On September 24, 2024, the announcement trailer for the upcoming DLC for Shift Up's *Stellar Blade* (2024) confirmed the much anticipated crossover collaboration with *N:A* (Playstation 2024). The next day, @NieRGame (2024), the official Twitter/X account for the *Drakengard/NieR* franchise, reposted the teaser trailer along with a short statement: "We will gladly say yes to anything for money." The post was broadly taken as a joke originating from *N:A*'s quirky auteur director Yoko. At the same time, though, it speaks to the afterlife of *N:A* within a gaming industry and culture that increasingly revolves around recurrent user spending.

---

2   "Though their exchange value may be complicated by emerging digital economies, video games are none-the-less saleable objects, with equally complicated and obscured use and labour values, whose existence – particularly in terms of Benjamin's vision of the ever-present phantasmagoria of modernity – implicitly facilitate the ongoing operation of unequal exchanges, and unfulfilled political possibilities, and illusory dream-states that are the very conditions under which the fetishized spectacle that Benjamin and Debord equally reject is able to flourish." (Fraser 2016: 190).

Collaborations like the one between *Stellar Blade* and *N:A*, in which the intellectual property of one franchise shows up in another, have become quite popular and profitable. The host franchise often pays for permission to sell discrete packets of content from the collaborating franchise as a way to tap into its established fanbase. Yoko and publisher Square Enix have been particularly successful in pursuing *N:A* collaborations, having participated in over twenty to date ("Crossovers" 2024). Some of these collaborations include missions and story content, extending the narrative universe of one or both franchises with varying success and substance. The majority, however, simply transport 2B (and sometimes 9S and A2) into the host's fictional universe as playable characters or skin, alienated from their original narrative context and characterization.

Such is the case with the *Stellar Blade* collaboration. The two games share obvious similarities. Both feature sexy android protagonists with drone companions, deployed in stylish, acrobatic combat to fight on behalf of humanity in a post-apocalyptic setting. In a joint interview with Yoko, Hyung-Tae Kim, creative director of *Stellar Blade*, explicitly cited *N:A* as inspiration (Krabbe 2024). Kim praises Yoko as an "exceptional" storyteller, acknowledging "I can't do anything similar." Kim's admiration for Yoko's storytelling might suggest that their collaboration might include some kind of narrative development. In the end, though, the resulting crossover includes no new mission or narrative content, only a collection of *N:A* themed cosmetics for Eve purchased from Emil's shop.

The promiscuous afterlife of *N:A* recalls Tsing's definition of alienation. In the case of *N:A* one might say the complex characterization, profound narrative, and empathetic gameplay would be the "ruins" that remain once the game's recognizable character art is extracted. While the prospect of reconnecting with *N:A* characters might bring some renewed excitement to fighting yet another Old Droid in the Great Desert, alienating them from their original narrative contexts risks diluting their significance as they become mere mobile assets. At the same time, as the work of Benjamin (2002) emphasizes, even commercial products can constitute an "image [...] wherein what has been comes together in a flash with the now to form a constellation" (463).

If the ruins of gaming include the alienating conversion of play and players into mobile assets, then *N:A* and its crossover afterlife demonstrate the value of creative reading and creative play. For Fredric Jameson, "the presence of some utopian content" can be found in even "the most degraded and degrading type of commercial product," if we are willing "to maintain that everything is a figure of Hope" (qtd. in Wegner 2024). The point here is not to disparage *N:A* or *Stellar Blade* as "degraded and degrading," despite their role in perpetuating sexist tropes and stereotypes (cf. Gach 2024). Rather, the aim is to underscore, following Tsing (2017), that "in the global state of precarity, we don't have choices other than looking for life in this ruin" (6). Not every game can possess the complexity or empathy of *N:A*, nor is it guaranteed these qualities will persist as intellectual properties are divided up, repackaged, and sold. Even so, the incorporation of *N:A*'s characters into other games like *Stellar Blade* invites players to (re)explore the themes of both, opening the possibility for moments of recognition and resonance. Perhaps these, too, are fragile instances of inflection, where the potential of ruins to disrupt narratives of progress leads players to reflect on surviving precarity.

References

"Crossovers." (6 November 2024) In: NIER Wiki | Fandom. Retrieved from https://nier.fandom.com/wiki/Crossovers.

"NIER Wiki | Fandom" (11 November 2024). Retrieved from https://nier.fandom.com/wiki/NieR_Wiki.

"The Ark: NieR Automata Lore," accessed November 11, 2024, https://theark.wiki/w/Welcome

2K Boston (2007): Bioshock. 2K Games.

Ash, J. (2012): "Attention, Videogames and the Retentional Economies of Affective Amplification," Theory, Culture & Society 29 (6), pp. 3–26, https://doi.org/10.1177/0263276412438595.

Benjamin, W. (2002): The Arcades Project, trans. Howard Eiland and Kevin McLaughlin. Cambridge: Belknap Press.

Bulut, E. (2013): "Seeing and Playing as Labor: Toward a Visual Materialist Pedagogy of Video Games Through Walter Benjamin." Review of Education, Pedagogy, and Cultural Studies 35 (5), pp. 408-425.

Bulut, E. (2020): A Precarious Game: The Illusion of Dream Jobs in the Video Game Industry. Ithaca: Cornell University Press.

Cavia. (2010): NieR. Square Enix.

Clemps. (2015-2018): "Drakengard Analysis." YouTube playlist, Retrieved from https://www.youtube.com/playlist?list=PLI1L9JVl6TFdwfvKUuebNp_zOAL5XXDT5.

Dookeran, J. (2024): "I Stopped Playing Video Games That Don't Respect My Time (You Should Too)," In: How-To Geek, 1 October. Retrieved from https://www.howtogeek.com/i-stopped-playing-video-games-that-dont-respect-my-time-you-should-too/.

Dyer-Witheford, N./De Peuter, G. (2006): "'EA Spouse' and the Crisis of Video Game Labour: Enjoyment, Exclusion, Exploitation, Exodus." Canadian Journal of Communication 31 (3), pp. 599–617.

Fraser, E. "Awakening in Ruins: The Virtual Spectacle of the End of the City in Video Games." Journal of Gaming & Virtual Worlds 8 (June 1, 2016): 177–96, https://doi.org/10.1386/jgvw.8.2.177_1.

Fuchs, M. (2014): "Gamification as Twenty-First-Century Ideology." Journal of Gaming & Virtual Worlds 6, 1 June.

Gach, E. (2024): "The Stellar Blade 'Censorship' Circus Explained." In: Kotaku, 29 April. Retrieved from https://kotaku.com/stellar-blade-censorship-eve-outfit-ps5-patch-1851442841.

Gamasutra Staff. (2017): "Does Visceral's Closure Prove AAA Single-Player Games Are Dying?" In: Game Developer (Formerly Gamasutra), 18 October. Retrieved from https://www.gamedeveloper.com/business/does-visceral-s-closure-prove-aaa-single-player-games-are-dying-.

Gerrish, G. (2018): NieR (de) automata: Defamiliarization and the poetic revolution of NieR: Automata. In: Proceedings of Nordic DiGRA 2018 Conference, 1 Jan.

Haines, C. (2021): "Nier: Automata and The Future of Planet Earth" In: Gamers with Glasses, 24 April. Retrieved from https://www.gamerswithglasses.com/features/nier-automata-earthfutures.

Hayes, C. (2025): The Sirens' Call: How Attention Became the World's Most Endangered Resource. New York: Penguin Press.

Hocking, C. (2007): "Ludonarrative Dissonance in Bioshock," In: Click Nothing, 7 October. Retrieved from https://www.clicknothing.com/click_nothing/2007/10/ludonarrative-d.html.

Irrational Games (2013): Bioshock: Infinite. 2K Games.

Jacevic, M. (2017): "'This. Cannot. Continue.' -- Ludoethical Tension in NieR: Automata." Presented at The Philosophy of Computer Games Conference, Krakow, Poland, 29 November. Retrieved from https://gamephilosophy2017.wordpress.com/wp-content/uploads/2017/11/jacevic_pocg2017.pdf

Jagoda, P. (2020): Experimental Games: Critique, Play, and Design in the Age of Gamification. Chicago: University of Chicago Press.

Janik, J. (2019): "Ghosts of the Present Past: Spectrality in the Video Game Object." Journal of the Philosophy of Games 2(1). Retrieved from https://ruj.uj.edu.pl/entities/publication/c12e86b3-54ae-42b1-bdfe-6f66dc1e746d.

Jenkins, H. (2006): Convergence Culture: Where Old and New Media Collide. New York: New York University Press.

Joseph-Gabriel, D. (2015): "Your Phone, Coltan and the Business Case for Innovative Sustainable Alternatives," In: Humanity in Action, February. Retrieved from https://humanityinaction.org/knowledge_detail/your-phone-coltan-and-the-business-case-for-innovative-sustainable-alternatives/?lang=nl.

Juul, J. (2012): A Casual Revolution: Reinventing Video Games and Their Players. Cambridge, MA: The MIT Press.

Kline, S./Dyer-Witheford, N./De Peuter, G. (2003): Digital Play: The Interaction of Technology, Culture, and Marketing. Montreal: McGill-Queen's University Press.

Krabbe, E. (2024): "Stellar Blade X NieR: Automata: Yoko Taro and Hyung-Tae Kim on How Their Blockbusters Inspire One Another." In: IGN, 15 April. Retrieved from https://www.ign.com/articles/stellar-blade-x-nier-automata-taro-yoko-hyung-tae-kim.

Kücklich, J. (2005): "Precarious Playbour: Modders and the Digital Games Industry." The Fibreculture Journal 5. Retrieved from http://five.fibreculturejournal.org/fcj-025-precarious-playbour-modders-and-the-digital-games-industry/.

Lasker, J. (2008): "Inside Africa's PlayStation War." In: Toward Freedom, 8 July. Retrieved from https://towardfreedom.org/story/archives/africa-archives/inside-africas-playstation-war/.

Lima, L.B./Walesa, D. (2022): "The Dark Play of Monstrosity in NieR: Automata." Journal of Gaming & Virtual Worlds. Retrieved from https://doi.org/10.1386/jgvw_00059_1.

Mills, E./Bourassa, N./Rainer, L./Mai, J./Shehabi, A./Mills, N. (2019): "Toward Greener Gaming: Estimating National Energy Use and Energy Efficiency Potential," The Computer Games Journal 8 (3), pp. 157–78.

Möring, S./Leino, O. (2016): "Beyond Games as Political Education – Neo-Liberalism in the Contemporary Computer Game Form." Journal of Gaming & Virtual Worlds 8 (2), pp. 145–61.

Murphy, D.T (2016): "Hybrid Moments: Using Ludonarrative Dissonance for Political Critique." Loading… 10(15). Retrieved from https://journals.sfu.ca/loading/index.php/loading/article/view/147.

NieR Series [@NieRGame]. (2024): "We Will Gladly Say Yes to Anything for Money." In: X (formerly Twitter), 25 September. Retrieved from https://x.com/NieRGame/status/1838951539097915539

Odell, J. (2020): How to Do Nothing: Resisting the Attention Economy. Brooklyn, NY: Melville House.

Paquet, A (2021): "The Automata Collective: Negation of Endings and Collective Formation in Nier: Automata." Mechademia 14(1).

Parker, F. (2018): "Roger Ebert and the Games-as-Art Debate," Cinema Journal 57 (3), pp.77–100.

Paul, C.A. (2024): Optimizing Play: Why Theorycrafting Breaks Games and How to Fix It. Cambridge, MA: The MIT Press.

Pérez-Latorre, Ö./Oliva, M. (2019): "Video Games, Dystopia, and Neoliberalism: The Case of BioShock Infinite." Games and Culture. 14 (7–8). https://doi.org/10.1177/1555412017727226.

Pérez-Latorre, Ó. (2019): "Post-Apocalyptic Games, Heroism and the Great Recession." Game Studies 19 (3). https://gamestudies.org/1903/articles/perezlatorre.

Platinum Games (2017): Nier:Automata. Square Enix.

Playstation (2024): "Stellar Blade – NieR: Automata DLC & Updates | PS5 Games." In: YouTube, 24 September. Retrieved from https://www.youtube.com/watch?v=6Xd9XNr6PAM.

Robinson, M. (2017): "Nier: Automata Was Almost a Farmville-Style Mobile Game." In: Eurogamer, 3 March. Retrieved from https://www.eurogamer.net/articles/2017-03-03-nier-automata-began-life-as-a-mobile-game.

Sarkar, S. (2017): "EA's Star Wars 'Pivot' Is a Vote of No Confidence in Single-Player Games." In: Polygon, 18 October. Retrieved from https://www.polygon.com/2017/10/18/16491188/ea-star-wars-visceral-games-single-player.

Schell, J. (2010): When Games Invade Real Life. Presented at DICE Summit, Las Vegas. Retrieved from https://www.ted.com/talks/jesse_schell_when_games_invade_real_life.

ShiftUp (2024): Stellar Blade. Sony Interactive Entertainment.

Smith, C. (2024): "Of Cyborgs and Cats: Nonhuman Companionship and the Specter of Humanity in NieR: Automata and Stray." Journal of Games Criticism 6, 27 October 27. Retrieved from https://gamescriticism.org/2024/10/27/smith-6-a/.

Soderman, B. (2021): Against Flow: Video Games and the Flowing Subject. Cambridge, MA: The MIT Press.

Solberg, R. (2021): "Playing Posthumanism? NieR: Automata and the Inescapable Human." Presented at Society for Literature, Science, and the Arts – EU Conference, Bergen, Norway. Retrieved from https://doi.org/10.33767/osf.io/crz8g.

Sun, Y. (2022): "Digitalising Labour By Attention Economy In Online Games." International Journal of Arts Humanities and Social Sciences Studies 7 (9). Retrieved from https://www.ijahss.com/Paper/07092022/1179451701.pdf.

Tanaka, M. (2014): Apocalypse in Contemporary Japanese Science Fiction. New York: Palgrave Macmillan US.

Tsing, A. (2017): The Mushroom at the End of the World: On the Possibility of Life in Capitalist Ruins. Princeton, NJ: Princeton University Press.

Wahba, M. (2018): "Nier and Nietzsche: Conveying Nihilism Through Mechanics." In: Scholarly Gamers, 27 June. Retrieved from https://scholarlygamers.com/feature/2018/06/27/nier-nietzsche-conveying-nihilism-mechanics/.

Wegner, P.E. (2020): Invoking Hope: Theory and Utopia in Dark Times. Minneapolis, MN: Univ Of Minnesota Press.

Wysocki, M./Schandler, M. (2013): "Would You Kindly? BioShock and the Question of Control." In: Ctrl-Alt-Play: Essays on Control in Video Gaming, edited by Wysocki,M. New York: McFarland, pp. 196-207.

Yablon, N. (2010): Untimely Ruins: An Archaeology of American Urban Modernity, 1819-1919. Chicago: University of Chicago Press.

Yeates, R. (2021): American Cities in Post-Apocalyptic Science Fiction. London: UCL Press.

Zhou, Z. (2022): "When Machines Long for Human Warmth : Nier: Automata and the Player-Game Relationship." MA Thesis, University of British Columbia. Retrieved from https://doi.org/10.14288/1.0417480.

"This Whole Place is Built from Ghosts"
Playfully Imagining Community and Hope in the Ruins of Capitalism in *Citizen Sleeper*

Ian Sturrock

Abstract

The marketing situates Citizen Sleeper *in the "ruins of interstellar capitalism", emphasising the remoteness and lawlessness of its setting. As Dyer-Witheford and de Peuter (2009) explain, the capitalist "Empire" is so pervasive that subversive play can only occur at its margins, exactly where the game is placed. In this paper, I close read (Bizzocchi/Tanenbaum 2011)* Citizen Sleeper, *using lenses of empire, capitalism, adaptation, and ruin.*

The game begins in ruin: the aforementioned ruins of capitalism, and the ruin of the player-character, whose body is physically breaking down due to obsolescence, planned by capitalism. The space station setting was made by capitalists but has been appropriated by post-capitalist factions: "We are keeping this place alive, but also remaking it into something new, dragging it away from those corporate origins," as Feng, the Systems Engineer, puts it in game. Work on the station is also ruinous, involving dismantling damaged spacecraft for parts and materials, but also a constant process of negotiation.

Surviving ruin by adaptation, mitigation, negotiation, and cooperation – this has a clear relevance to our own apocalyptic present and future. In particular, this relates to Bendell's writing (2020) on deep adaptation, and to Brooks and Agosta's work (2024) on the evolutionary biological aspects of humanity's responses to the climate crisis. These researchers show that neither a technological, nor a political, fix to the crisis is plausible; attempted fixes are doomed due to coming from a neoliberal capitalist paradigm.

Collapse and degrowth are coming, but they will not come from a plan made by corporate or civil authorities. Without such guidance, people must turn to games and stories to think about and try potential strategies to live through apocalypse. Reading Citizen Sleeper *will allow us to reverse-engineer patterns (Björk/Holopainen 2005) that can be used by game designers to create games to guide humanity in adaptation to the crises of the Anthropocene.*

Keywords

ruin, game design, deep adaptation, evolutionary biology, empire, climate crisis, ecological crisis

Introduction

"We learned not to talk of hope on a dying world. There is meaning for them, and any future they can build with the time they have."

AKI, IN CITIZEN SLEEPER, 2022

Citizen Sleeper is a critically acclaimed game, with a Metascore of 82 (Metacritic 2025). It's remarkable for its conscious focus on solidarity, nurturing, and community as the primary ways to navigate through its dystopian cyberpunk future. Farca points out that dystopian games commonly have threads of hope: assorted ways in which the player can somewhat fulfil regenerative and even utopian impulses, despite the devastation of societal collapse (2018), but it's rare to come across one that foregrounds the utopian potential at the end of capitalism to such a degree. It "breathes a vibe of connection against the darkness, of fragile life emerging through ecologies of care" (Vervoort 2023). It subverts conventional game design tropes of empire-building and combat to craft an opposition to capitalism and violence through kindness and fellowship.

This consciously adversarial response to existing videogames – especially games that take the visual style and surface-level tropes of cyberpunk and other speculative fiction, without paying more than lip service to the anti-authoritarian themes of those settings – makes for powerful, immersive, and unusually hopeful ludic storytelling. This can help its players conceive of, and perhaps begin to create, alternatives to capitalism and violence, even outside the game world.

The game starts out with familiar tropes: a cyborg protagonist; a decaying space habitat with an angry, exploited, and marginalised population; bounty hunters and corporate assassins and mercenaries. It foregrounds the radical and speculative aspects of each to give the player not just the sense of agency that any good videogame will, but a sense that the player might be able to initiate new ways of resisting oppression, new ways of living, new and more egalitarian communities.

Sleeper: a cyborg player-character

Citizen Sleeper opens with crisis and desolation: you play the role of a kind of cyborg known as a sleeper, with a partially organic but artificial body that acts as the hardware for an emulated mind scanned from the brain of a still-living human. In this setting, sleepers were created as disposable workers by unscrupulous corporations, and are seen as the property of their creating corporation. They are capable of performing tasks and operating in environments that would be impossible or dangerous for humans, and of course as property, they are given neither pay, nor rights.

There are clear echoes here of the replicants in *Blade Runner* or the androids in *Do Androids Dream of Electric Sheep?*, which likewise are primarily organic yet artificial creations, somewhat more physically robust than humans but seen as property; and, of course, echoes of the real-world institution of slavery that are also a well-supported reading of the theme of artificial humans in those stories, by for example Bertek (2014). Indeed, artificial beings as either metaphorical or literal slaves of society, is a common science fictional theme dating back to Čapek's *R.U.R.* (1920), the stage play for which the term "robot" was invented; the robots of *R.U.R.* are also created organically and are human-like in appearance, and carry out a successful rebellion against their human creators and owners.

The notion of emulating specific, living human personalities in an artificial form is also common in literary science fiction, with for example the *moravecs* of Simmons's *Ilium* (2003). This concept is also present in speculative but technical scientific papers such as *The logical core architecture* (McKendree 1998) in which human personalities are proposed to be archived in tiny but infinitely expandable nanotechnological spacecraft; essentially compressing a skilled human worker into a file that can then be inserted into a body that is built "to order" in situ on an exoplanet, reducing the need to physically carry humans into space so as to colonise the galaxy cheaply and efficiently.

The player-character in *Citizen Sleeper* does not have a name, being addressed simply as "Sleeper" by other characters in the setting. They have little to no memory of their life prior to arriving on the Erlin's Eye space station; neither their immediate past working for the Essen-Arp corporation as a sleeper, prior to escaping, nor their personality's previous existence in a human body. This literary motif of amnesia is common in videogames, particularly computer roleplaying games such as *Planescape: Torment* (1999) and *Disco Elysium* (2019), both of which use the motif to create "structures of discontinuity and rupture in the represented ludic self" (Vella/Cielecka, 2021). An amnesiac main character also allows game developers to craft an emotionally powerful narrative in which discoveries and revelations about the player-character's past can be introduced at suitably dramatic moments; the critically acclaimed *Disco Elysium* (2019), perhaps the most obvious CRPG influence on *Citizen Sleeper*, makes full use of this affordance.

Further, the amnesia motif allows the player to more easily empathise with, and project their own choice of personality onto, the player-character, which is a major use of it in *Citizen Sleeper*. In this case, the amnesia also works to enhance the vulnerability of the player-character. Not only are you not a mighty hero, festooned with guns, but you are reliant on the kindness of strangers, a refugee who has left absolutely everything behind including your own memories and your past.

As a cyborg, Sleeper is inherently a somewhat contradictory, ambiguous, transhuman figure. Neither fully human, nor fully machine, they perfectly represent humanity on the cusp of becoming posthuman, thus enhancing the game's sense of being at a pivotal moment: potentially the end of capitalism, potentially the beginning of something new. They are physically more enduring than humans, but this comes at the cost of reliance on medication, a "stabiliser" drug originally manufactured by the corporation that built Sleeper, Essen-Arp, but potentially available from other sources during play. Finding a source for stabiliser, and a source for the often significant cost in credits (the currency on *Erlin's Eye*), is an ongoing task and one that can create some level of anxiety. This works well as a metaphor for the reliance on corporate healthcare in the early 21st century, an era in which even in countries with socialised healthcare, there is often a cost to medication as well as supply chain issues that can leave us unable to acquire it as easily as we might hope. Fictional heroes, particularly cyborgs, are rarely depicted as being quite so vulnerable, or so empathetic.

Cyborg status in science fiction typically involves a character who has suffered an injury, often during military labour, and has had the injured body part(s) replaced with prosthetics that not only perfectly recreate the functionality of the original, but enhance it in some way: see, for example, the portrayal of Geordi La Forge in *Star Trek: The Next Generation*. His 'VISOR' vision replacement not only cures his blindness, but gives him the capacity to perceive far beyond the human visual spectrum. As Wälivaara points out, "portrayals of futurescapes in which cures of disability are highlighted are closely connected to a medical paradigm of disability" (2018). Utopian futures depict disabled characters only as 'cured' or even enhanced compared to baseline humans.

Dystopian futures show disabled characters as an indication of that dystopia: "the disabled body has come to signify not having a future or that the future has failed" (Wälivaara 2018). While the depiction of disabled characters in *Citizen Sleeper* might be interpreted in this light, it could be argued that this depiction is at least as much about having representation at all: other than Sleeper, there are three characters with some kind of visible, or textually referred to, disability: one who uses a mobility aid due to age, and the other two affected directly by the manipulations of corporate entities that treated them primarily as assets rather than people. Could this be, at last, what Wälivaara argues as the ideal, "a sign of a future society in which different types of bodies are embraced" (2018)? Certainly the sense of solidarity and community in *Citizen Sleeper*, argued throughout this work to be a major theme, is absolutely inclusive of marginalised bodies of all kinds. What unites all the non-player characters (NPCs), and Sleeper themself, is a shared history of having suffered under capitalism, whether physically, mentally, financially, or otherwise; and a blurry but shared vision of a better and more egalitarian future.

This is not, after all, merely a dystopian fable, but one in which altruism is real and hope is worthwhile. Everyone has suffered, everyone has been exploited, but all have a part to play in planning and creating a new society.

Perhaps the one criticism one might make of the radical inclusivity of *Citizen Sleeper* is that, in including a variety of marginalised NPCs, whether disabled, BIPOC, poor/underclass, queer and/or non-binary, it ignores some of the specific ways in which such groups are marginalised in contemporary times by having all of them be similarly marginalised in-universe; potentially close to claims of "I don't see colour" uttered by conservatives as though to suggest that there are no social inequalities and thus no work must be done to address those inequalities. Still, in the game's marginal, liminal setting, at the very physical (and even astrophysical) limits of the power of capitalism, in what feels like the space-age equivalent of a sacrifice zone, it's no surprise that intersectionally marginalised populations will be found, and will work together, whether out of necessity, a sense of solidarity, or both. And, certainly, the sense of injustice, of work to be done, of changes that must come, can be found in almost every word or image in the game.

"You realise now that you are no different than anyone else on this station in the eyes of the people in that room, and those like them. Each body here can be recast as a piece of property, a tool, an expense, an acceptable loss." – *Citizen Sleeper*, 2022

Community and solidarity, in defiance of Empire

> "People don't leave their homes for nothing"
> SLEEPER, IN CITIZEN SLEEPER, 2022

Sleeper can, and indeed must, build relationships with various NPCs on *Erlin's Eye*, some of whom are sympathetic to them and their status as a penniless refugee, while others seek only to abuse or exploit them. By situating the player-character with no resources or friends to begin with, the game echoes a standard CRPG and TTRPG "character advancement" path, which starts characters off at Level 1 and anticipates they will progress ever upward; but *Citizen Sleeper* also subverts the pseudo-capitalistic, empire-building or engine-building qualities of traditional character advancement, which tend to focus on rapidly acquiring and upgrading weapons, armour, and related equipment.

Jayanth points out that the "gun is still the fundamental tool we give players to interact with the world", advocating instead for games in which our fundamental human needs "for identity, community, self-esteem, challenge, love, joy" are fulfilled through the game, rather than deferred (2021). *Citizen Sleeper* is, perhaps, at last, one such game.

By contrast to the usual weapon-focused engine-builder, the game uses Sleeper's lowly status to support the player's empathy with outsiders, those systemically disadvantaged and Othered by mainstream society: refugees, the disabled community, the queer community, gig-economy workers and the underclass. In *Dungeons and Dragons* (Arneson/Gygax, 1974; hereafter *D&D*), the first tabletop role-playing game, character advancement is achieved partly by acquiring gold pieces to enable you to "level up", that is, acquire the aforementioned combat improvements; whereas in *Citizen Sleeper*, money (credits) is much more likely to be used to stay alive, or to assist other members of Sleeper's found family and community.

Conventional videogames in most genres, not just CRPGs, afford the player two significant tools with which to affect the game-world and achieve their goals: combat and resource management. There may well be other tools: perhaps stealth, persuasion, or other game social mechanics. But combat and resource management are almost universal.

These two, even in games that purport to oppose capitalism, are very much normative of capitalism and of Empire; they are the standard tools that have always been used in the real world to build and maintain hegemonic power at the expense of others. Resource management in such games tends to involve building an "engine" or power-base; stacking various bonuses, buildings, or gizmos together in-game so as to be essentially unbeatable, and then using that to dominate the game, usually either combatively or economically. Again, this feels like an unintended metaphor for capitalism as it tends to operate: not with the invisible hand regulating all, as per the dreams of libertarians, but rapidly degenerating into monopoly, market dominance, militarism, or totalitarianism, once sufficient resources have been concentrated into a small enough number of hands.

Violence in *Citizen Sleeper* is not a tool; it is something inflicted on you and those you love, something to be endured and survived, if you are lucky. Violence is a part of life in the ruins of capitalism, even less orderly and predictable than the violence of capitalism at its full height. There are a handful of scenes in which Sleeper sees or encounters violence, and they're all chaotic and terrifying, whether being present at a riot where disgruntled workers realise their bosses have cheated them, or being attacked by armed professionals and desperately making panicked decisions in the hope of living through to the next cycle. This is a rare game in which violence feels real not because of a meticulous attention to ballistics and other statistics, but because of the emotional intensity of its depiction; rarer still to recognise that violence is primarily the tool of the oppressor, not the liberator. That's not to say that the game entirely shies away from the notion of arming the underclasses and the refugees, but it does recognise that even righteously justified anti-oppressive violent intent often has unintended consequences and unintended victims.

There is resource management here, too, but there is no engine to build, no capital assets to acquire, no optimisation of a personalised Empire with which to force the universe to obey your will. Rather, your resources are scant and limited. In a design decision most likely inspired by contemporary Euro-style hobby boardgames rather than other videogames, you will never have enough capacity to do everything you might want to do, in time. As a part of the underclass yourself, you don't have spare resources to hoard so as to become a part of the capitalist class; you can barely keep yourself alive and functional, at least in the early stages of the game.

The main resource you have to begin with is your labour; your time and your hard-working body. This is represented by action dice: standard, six-sided, virtual dice, between one and five in number, which the game rolls for you at the start of the cycle (the game's term for a space station day) and which you may then assign to actions.

Again, this feature is clearly inspired by contemporary non-digital games, notably worker placement games and their descendants the dice worker placement games – see, particularly, *Kingsburg* (2007) as an early example of the latter, but closely related mechanics have been used in everything from miniatures wargaming such as *Saga* (2011), to heavily thematic, narrative board games such as *Dead of Winter* (2014). Worker placement games involve the player being given a number of pawns or other tokens in each round of play, and physically placing them, one at a time, on specific spaces on the gameboard, each representing a worker in the game's setting, taking a particular action. For example, in *Agricola* (2007), one begins with two pawns, representing the members of a subsistence farming family in medieval Germany, and might place one of them on the Sheep space, representing that worker spending some time to acquire additional sheep for the farm. The dice worker placement game mechanic builds on this concept, but in this case, the available dice are typically rolled at the start of a game's round, and so provide an additional decision point: the value on each die will have an impact in how effective a given action is.

Worker placement games often have a rather capitalistic quality: the player is putatively in a position of authority over a number of workers, and can thus assign them to various tasks, usually again with an intention to build up further resources and optimise one's in-game position. By the time they evolve into *Citizen Sleeper*, though, it's essentially the labour of one individual that one is controlling, the player-character, Sleeper themself. And again, there is no empire to build here; without some labour devoted to simple survival, the acquisition of credits to buy food and medication, Sleeper's existing bodily resources will dwindle, and they will find they have fewer dice to use in any given cycle. Some actions will carry a risk of negative consequences, particularly if the player assigns low-value dice to them; perhaps Sleeper will be wounded, or if they are performing an illicit activity, an authority figure might be alerted to their actions. The majority of actions represent labour of some kind, often a shift dismantling a ruined starship, scavenging for scrap in a disused part of the station, or perhaps working for a wider community such as the egalitarian Hypha Commune or the vulnerable refugees of the Flotilla.

Most actions involve clicking and dragging a dice from Sleeper's dice pool to a dice space on the relevant action on the screen, or sometimes instead, clicking and dragging a different resource such as a piece of acquired data or a chunk of scrap. In each case there will be a short delay before seeing the results of the action, a delay of just a second or two in which the player has nothing to do other than watch the screen and wait for the action bar to fill up.

As a player, making use of these actions and then having to wait for their results, rapidly becomes slightly tedious. It's only a delay of a second, but if, for example, your previous computer hacking activities have netted you six or seven pieces of data on the Yatagan gang, and you wish to sell them to the gang's enemies, Havenage, you will need to perform six or seven clicks and drags, and wait for six or seven action bars to fill up. Partway through play, though, I began to see this minor grinding- or farming-type of activity as an extension of the labour metaphor: just as Sleeper arrives on *Erlin's Eye* with no resources other than their own body and mind, and must make use of those resources to achieve anything, including survival, so the player must devote their own time and effort to represent that same labour on Sleeper's part. Yes, it is potentially boring, but only for seconds; a small reflection of the travail and danger that Sleeper undergoes in a full work shift spent cutting through a starship wreck.

Where in a typical Empire-style game, the player must assemble various resources to create more powerful resources and indeed capital assets – for example, acquiring wood and stone to craft a marketplace that can then be used to generate passive income and other benefits – Sleeper rarely gets a chance to hold on to resources for long, and never to create a capital asset. The closest thing to this is perhaps their potential rejuvenation of the Aviary, an in-game ruined location that can be cleaned up and repurposed as a fungus farm, but even then, it requires regular labour on Sleeper's part to keep it functional, and the various spores and mushrooms that can be acquired there rapidly find uses, feeding oneself or helping one's community and found family. Again there is a significant difference in the symbolism of this as compared to most games; rather than having the newfound capability to impose their will on more and more fictional workers, territories, troops, starships, and so on, Sleeper has a newfound capability to assist those who need it the most, to repay past kindnesses from the time when they were a vulnerable refugee, to pay it forward and aid those who are similarly at risk now. A truism in ecoactivism circles is "We are all climate refugees now" (Sachs, 2018). Being a refugee, being helped, and then helping other refugees, is a cycle. Genuinely resilient communities must also be altruistic and welcoming communities: not hostile or conquering forces.

A gradual shift in the game's significant resources, from purloined data as the most crucial one in the early game, to carefully cultivated fungi in the endgame, represents a shift from the classic cyberpunk tropes of the outlaw hacker, to something new, a sense of being a part of a larger community, the working-class and underclass community who make up most of the population of *Erlin's Eye*. The fungi, in all their weird and wonderful varieties, whether grown from spores by Sleeper in the Aviary, or subversively spread throughout the managed *Erlin's Eye* biosphere as part of a wider effort to oppose a corporate power, represent more than just solidarity and community. They also embody hope for the future.

Mushrooms are important in the real world, acting not just as food, medicine, psychoactive drugs, and other uses, but also having a cultural importance: ethnobotanists classify them according both to "use value" and "Index of Cultural Significance" (ICS), recognising their vital role across many cultures. In transplanting real-world mushrooms into its ruined space station setting, adding some new ones thanks to a similarly ruined AI gardener, and then giving the player agency to further cultivate and spread these fungi in direct opposition to the corporate forces that would take over *Erlin's Eye* again, *Citizen Sleeper* offers a thread of hope leading from Indigenous and other cultures on Earth, through the capitalist hell of the present and near future, into a potential new, post-capitalist reality, in which traditional values and cultures offer us insights into how to evolve new values and cultures. As Tsing points out, "mushroom collecting brings us[...] to the unruly edges and seams of imperial space, where we cannot ignore the interspecies interdependencies that give us life" (2012); the most culturally significant and useful fungal species are found in liminal spaces, borderlands, and the very fringes of capitalism, precisely where *Citizen Sleeper* is set. As neither a plant, nor an animal, and as both a giver of nourishment and health, and a powerful emblem of poison and death, the mushroom occupies a zone of ambiguity in much the same way as the cyborg.

The shift in resource importance, from the data of the first half of the game to the fungi of the second half, also reflects two shifts in tone, one relating to Sleeper's place in *Erlin's Eye*, and one relating to the wider setting of the game. Initially, Sleeper, as a penniless and technically homeless refugee, needs data primarily to sell to various contacts, to gain credits, to stay alive. By the time of the shift to a "mushroom economy," not only has Sleeper acquired at least one semi-permanent place to live – without having to rely on the sufferance of a kindly NPC – but they are also very much a part of the *Erlin's Eye* community, with contacts, friends, colleagues, and even found family, all over the station, from bars to docks, most of them well-disposed to the player-character because they know Sleeper to be a reliable, hard-working comrade and kindly friend. And Sleeper is no longer reliant on credits just to stay alive: other options have opened up, including communal living, or preparing one's own mushroom dishes for food. Instead, they can use their newfound mushroom resources to help the wider community, whether feeding refugees, enhancing the offerings at their friends bars and food stands, or aiding the botanical research of the Hypha Commune.

"Eshe, we need help. Just like the refugees needed us, we need others too." – Peake, in *Citizen Sleeper*, 2022

Roleplaying in the ruins

> "an… atrium, still lit by some ragged looking light panels on the ceiling, once simulating a summer sky, now degraded into a flickering blue and white tapestry of glitches"
>
> AREA DESCRIPTION, IN CITIZEN SLEEPER, 2022

D&D has influenced TTRPGs and CRPGs enormously, and continues to do so, setting the tone and theme for decades of imitators and remediators. Given its age and its origin in the USA, it is no surprise that *D&D* has a libertarian, frontier ethos running through both its system design and worldbuilding: beginning, low-level player-characters are expected to move from place to place like roving gunslingers, seeking fame and fortune, effectively outside the law. At higher levels, player-characters are frequently framed as taming the wilderness and founding new settlements. Though ruins abound in *D&D* adventures, they're portrayed primarily as another form of "dungeon": that is, places that adventurers travel to, so as to kill the new occupants and loot their valuables. At least in early iterations of *D&D*, there's rarely a social context to the ruin; we usually do not care who built it, or why it is ruined, or what became of the original inhabitants. Exceptions do exist but tend to be later developments and/or come from outside of the USA: Albie Fiore's *The Lychway* (1978), for example, exemplified and began a more ecologically rounded, complex tradition of dungeon design, in which the history of the ruin, and the culture of its former occupants, are materially important to solving the central puzzles of the story.

Citizen Sleeper takes place almost entirely in the inhabited, still active, ruinous world of a space station, the aforementioned *Erlin's Eye*. Again, the history of the place is important to the game's story, but more so is the mixture of continuity and fragmentation of the original community. Although some kind of collapse-enabled revolution overthrew the original capitalist, corporate control of the place, even changing its name from the original Æ1 or *Ash One*, many of the workers and the poor remained on the station, unable or unwilling to give up their home. Refugees and others have arrived since.

Significant elements of the original, elitist power structure remained in place, too, and corporations exterior to the station are able to send agents there and otherwise influence and interfere with life on *Erlin's Eye*, giving the game a cynically realist quality: capitalism may be in the process of self-destruction or even self-cannibalisation, but it retains the power to do harm, even in its death throes. The vast ruins of the *Eye* allow for all kinds of secrets: there is more than enough space here for elaborate, high-stakes plots, with rival corporations secreting military or industrial resources in supposedly empty sectors as part of takeover plans or other hostile acts.

Some of the spaces are ruined in much the same way as one might expect in poorer areas of 21st century cities: housing units that were never completed due to a lack of money, bars and cafes seemingly squatted or appropriated from buildings intended for other uses, whole regions and economies that are devoted to scavenging and repurposing technology and materials that were left behind after accidental or other destruction. Other sections of the *Eye* are ruined more in the way one might find in a horror story or some of the more mythologically influenced, classic cyberpunk literature such as *Count Zero* (Gibson 1986): here you will find ghosts in the machine, artificial intelligences with their own agendas after escaping corporate control, and hunter-killer software that is still deadly despite now being obsolete or purposeless. Others still are not simply ruined, but rewilded: a whole section of the *Eye*, the Greenway, is a former hydroponic garden, no longer under corporate control but overgrown, strange, and filled with all of the potential of life and evolution. The aesthetic pleasure in ruin-sentiment that Sleeper feels in the latter sections, reminds us that ruins are intersections between artifice and nature, as explored by Vella (2010):

"You try to imagine what this place looked like before the collapse. All ordered rows of crops, corporate pleasure gardens, glowing farm stacks. Now it is a wilderness, a strange overgrown biosphere bordered by the void of space. You smile, despite everything, it seems like an improvement." – *Citizen Sleeper* (2022)

Disastrous and apocalyptic fiction, from the beautifully over-the-top *Mad Max* films (1979-present) to something grittier like *The Road* (McCarthy 2006), tends to emphasise a breakdown of the social order, suggesting that we are all one bad day away from stabbing each other over a scavenged tin of dog food. This, too, is arguably a conservative mindset, even when the creators may intend to challenge such views; capitalist realism makes cynics of us all (Fisher 2009). There is an underlying assumption that "human nature" will always involve competition and strife, leading to brutal warlords enforcing a violent rule.

A more hopeful perspective might emphasise humanity's powerful capabilities of altruism, cooperation, negotiation, and compromise, and this is the perspective that is found in *Citizen Sleeper*. When a lone agent or bounty hunter shows up to violently enforce a corporation's will on the station, they will not find a compliant or even sympathetic populace. Rather, the player can rely on ordinary people, even strangers, to want to help Sleeper, clearly the underdog, as best they can, even if that means squaring up against an armed outsider. This is in stark contrast to the frontier expectations of those of us raised not on *D&D* but on that other mid-20th century American cultural export, the Western: surely the townspeople hide in meek terror when the desperado shows up, awaiting rescue from a hopefully more benevolent gunslinger.

Solidarity among marginalised and vulnerable communities can be found frequently in the real world, but more rarely in apocalyptic fiction. Consider Lesbians & Gays Support the Miners, the London-based movement that raised money for striking miners elsewhere in the UK in the 1980s (see, for example, the archival material referenced by Fisher 2014); the renowned capacity of working-class Glasgow communities to prevent the deportation of immigrants by simply filling the streets with the very physical solidarity of their own bodies (Mainwaring et al. 2020); hungry civilian inhabitants of besieged Sarajevo sharing seedlings and food to ensure nobody starved entirely (Gray 2019). This is not to claim that warlords, sectarian violence, and other brutalities will never be found in the event of societal collapse, but more to suggest that the overwhelming tendency of fictive collapses to focus on said brutalities, makes it difficult to even imagine that humans can be altruistic, kind, welcoming, and benevolent, even and perhaps especially in times of the most extreme privation and desperation, despite evidence of such prosocial behaviour in dire situations in the real world.

Strategies of escape and renewal

> "You... you are a part of this place."
> PEAKE, IN CITIZEN SLEEPER, 2022

Several endings are available in *Citizen Sleeper*. You can flee *Erlin's Eye* in three different ways: with your newfound friends, with your found family, or with the refugee flotilla. Alternatively, you can practice what Haraway calls "staying with the trouble":

"Staying with the trouble means making oddkin; that is, we require each other in unexpected collaborations and combinations, in hot compost piles. We become – with each other or not at all." – Haraway, 2016

Escape is always tempting, and is, arguably, sometimes necessary for survival; as a climate crisis survival strategy, Brooks and Acosta argue that "Humanity must begin preparations for large-scale retreat and resettlement measures as soon as possible" (2024). Sometimes that means resettling more than once: as refugee flotilla captain Sol puts it, "I didn't bring my people here just to let them die somewhere new" (*Citizen Sleeper* 2022). There are good reasons in-game to believe that *Erlin's Eye* is ultimately doomed, either due to its own ongoing and worsening ruin, or due to the hostile takeover schemes of rival corporations. "This place has always been on borrowed time." – Peake, in *Citizen Sleeper*, 2022

The Celis Foundation purports to be using the *Eye* for essentially charitable purposes: to create a new, idealised, socialist society, even further beyond the reach of the capitalist Core. They claim to offer a lottery for inclusion in that new society, with a ticket for anyone who serves their time as a worker on their colony ship.

"Unlike most of the Core, we neither believe *Erlin's Eye* to be a threat or a rogue state, but instead an embryo for the formation of a new, decentralised social structure..." – the Celis Foundation, in *Citizen Sleeper* (2022)

But the inclusion of the Celis Foundation in the game is a metaphor for privileged colonialism, not for real altruism. Founding a new life by buying an old farm and turning it into a hippie commune, supposedly out of the reach of capitalism, is a bourgeois escape plan, not something for the working class or underclass: "You understand they never even put us on the list right? I've been all around the rim looking for work and I've run into more than a few from the crews. It turns out only long-time Havenage members were issued those Celis ID numbers, they never even planned to consider us." – Lem, in *Citizen Sleeper* (2022)

One of the escape-type endings involves getting aboard the Celis Foundation colony ship, but not through their kindness; Sleeper will have to turn their hand to hacking, forgery, and subversion once again if they are to acquire tickets for themself and their found family.

Corporate liberalism, and centrism in general, is depicted as dangerously compromised and complicit, as barely any different to the more hardline capitalist approach where every sentient being is just another asset. There's always a good reason to say "no, sorry, can't" to the obviously most ethical choices; who, after all, will pay to feed the refugees? Surely we must count the economic costs of altruism! "Can the Eye take in thousands of refugees safely? I don't know. This place is a ruin, Sleeper. What do you think the council discusses, the price of Girolle? We make plans to keep this place spinning, to keep life support working, to ensure..." – Councillor Helene, in *Citizen Sleeper*, 2022

Climate and social justice activists in the early 21st century follow Haraway's notion of "staying with the trouble" as best they can, seeing it as the most ethical option. Fleeing doesn't solve the wider problems; it doesn't save the wider community, just yourself and maybe a few others. You may have to go, but if you do, you will take your community with you. It's not made clear in-game as to whether you have a long-term future aboard the *Eye*, even if you can solve its most obvious problems; perhaps joining the refugee flotilla, and helping them take as many of the *Eye*'s community as they can, is the right choice. The game does not judge you either way, but there is a satisfaction in staying with the trouble, in knowing that you have given it your best shot to fix the trouble, you and your oddkin and your hot compost pile.

Oddkin, for Haraway (2016), is a very broad category, including not just our haphazardly discovered, sentient found family and community, but companion species too. Sleeper can feed a stray cat as part of their gradual integration into the society of *Erlin's Eye*, but the most obvious nonhuman companion species are the aforementioned mushrooms, probably even more essential to the future survival of the humans of the *Eye* than is Sleeper. This, of course, also fits the notion of the hot compost pile; one did not necessarily expect to be playing a mushroom farming and rewilding sim in one's futuristic cyborg CRPG, but it makes for a satisfying way to find some hope in a dystopian ruin.

The end of capitalism

> We live in capitalism, its power seems inescapable – but then, so did the divine right of kings. Any human power can be resisted and changed by human beings. Resistance and change often begin in art.
>
> (Le Guin 2014)

Capitalism will almost certainly fall in the 21st century, whatever else happens. Either governments finally enforce the Paris Agreement of 2015 and vastly reduce carbon emissions, an unlikely outcome given that it can't be achieved without economic degrowth, a fall in living standards, and the very likely dismantling of capitalism as we know it, none of which is likely to win over voters; or else, states and economic systems collapse entirely under wave after wave of worsening disasters, simply unable to pay for the level of rebuilding necessary as resources run out (Brons 2019), as well as being unable to maintain effective control over borders, a crucial way in which states are defined and ordered; borders will dissolve in a world with potentially 1.2 billion climate refugees on the move (Institute for Economics and Peace 2020).

The obvious criticism to make of my claim, above, is that I've only given two possibilities; surely life is not so binary. But the question of, do we meet the requirements of the Paris Agreement, or not, is a binary one. And, writing in 2025, with emissions targets failing to be met each year, with promise after promise broken, it's clear that even the usually cautious United Nations experts are admitting that only the most drastic emissions cuts will have a chance of staving off the most drastic of catastrophes.

What comes next? Either way, those of us who are fortunate enough to live through degrowth, or collapse, must find new ways to live, and perhaps new places, too, given how many of us, too, will become refugees. Games and other fiction that emphasise cooperation and kindness may help us to imagine how.

That's not to make the claim that games, even games as good as *Citizen Sleeper*, will allow us to somehow find the wherewithal to solve the climate crisis, or to force governments and corporations to solve it themselves, though Le Guin certainly suggests the latter as a possibility. Rather, it is to suggest that such forms of imaginative, interactive play are perhaps one of the easiest, most accessible, and most powerful ways to understand how to navigate through collapse, whether economic or civilisational or otherwise (Illingworth 2023). To remind us to stay with the trouble; to help us recognise that a hopeful future can only be found by working with the whole community, even those who might not immediately seem like our friends.

The counterargument is that digital games are an inherently energy- and resource-intensive mode of entertainment, intertwined with, and compromised by, the capitalist, extractivist system they are produced in, and so can't be trusted to point the way to an alternative to it. There is inarguably a lot of truth in all of this! I will point out that *Citizen Sleeper*'s running requirements are very low, allowing it to fit in with notions of low-carbon computing; and that this article, and other critique of games that strive to be radical, are a way of determining how much we can, indeed, trust attempts to depict alternatives to capitalism. As Jayanth points out, "Video games are only made possible by the kinds of technologies, knowledge-systems, collaborations, platforms, structures and even excesses of capitalism-colonialism;" but she also sees value in game designs that allow us to imagine alternative pleasures and satisfactions to those of capitalism, which Citizen Sleeper clearly does, even if it might not, quite, "make the revolution irresistible" (2021).

References

Bendell, J. (2020): *Deep Adaptation: A Map for Navigating Climate Tragedy*. Ambleside: Initiative for Leadership and Sustainability.

Bendell, J. (2023): *Breaking Together: A Freedom-Loving Response to Collapse*. Bristol: Schumacher Institute.

Bertek, T. (2014): 'The Authenticity of the Replica: A Post-Human Reading of Blade Runner'. *Literary Refractions*, No. 1, Year 5

Björk, S./Holopainen, J. (2005): *Patterns in Game Design*. Hingham: Charles River Media.

Black Isle Studios (1999): *Planescape: Torment*. Los Angeles: Interplay Entertainment

Blade Runner (1982): Directed by Ridley Scott.

Brons, L. (2019): 'A theory of disaster-driven societal collapse and how to prevent it'. *Knowledge Commons*.

Brooks, D./Agosta, S. (2024): *A Darwinian Survival Guide: Hope for the Twenty-First Century*. Cambridge: MIT Press

Buchel, A. (2011): *Saga*.

Čapek, K. (1920): *R.U.R.*

Carrington, D. (2024): "'Crunch time for real': UN says time for climate delays has run out", *The Guardian*. Available at: https://www.theguardian.com/environment/2024/oct/24/crunch-time-for-real-un-says-time-for-climate-delays-has-run-out (Accessed: 24th October 2024).

Chiarvesio, A./Iennaco, L. (2007): *Kingsburg*.

Dick, P. (1968): *Do Androids Dream of Electric Sheep?* New York City: Doubleday.

Dyer-Witheford, N./de Peuter, G. (2009): *Games of Empire: Global Capitalism and Video Games*. Minnesota: University of Minnesota Press.

Farca, G. (2018). *Playing Dystopia: Nightmarish Worlds in Video Games and the Player's Aesthetic Response*. Bielefeld: transcript Verlag.

Fiore, A. (1978): 'The Lichway'. *White Dwarf Magazine Issue 9*. London: Games Workshop

Fisher, E. (2014): 'Lesbians and Gays Support the Miners material at the People's History Museum', *People's History Museum*. Available at: https://phmmcr.wordpress.com/2014/09/29/lesbians-and-gays-support-the-miners-material-at-the-peoples-history-museum/ (Accessed: 24th October 2024).

Fisher, M. (2009): *Capitalist Realism*. London: Zero Books.

Garibay-Orijel, R./Caballero, J./Estrada-Torres, A./Cifuentes, J. (2007): 'Understanding cultural significance, the edible mushrooms case'. *Journal of Ethnobiology and Ethnomedicine*, Vol 3.

Gibson, W. (1986): *Count Zero*. Victor Gollancz: London.

Gilmour-Long, J./Vega, I. (2014): *Dead of Winter*

Gray, R. (2019): 'What happens when we run out of food?', *BBC Future*. Available at: https://www.bbc.com/future/article/20190319-what-happens-when-the-food-runs-out (Accessed: 24th October 2024).

Haraway, D. (2016): *Staying with the Trouble: Making Kin in the Chthulucene*. United States of America: Duke University Press.

Illingworth, S. (2023): 'How board games can get people involved in climate action', *The Conversation*. Available at: https://theconversation.com/how-board-games-can-get-people-involved-in-climate-action-209707 (Accessed: 7th November 2024)

Institute for Economics and Peace (2020): *Ecological Threat Register 2020*. Sydney.

Jayanth, M. (2021): "White Protagonism and Imperial Pleasures in Game Design #DIGRA21" *DIGRA India 2021 keynote transcript* . Available at: White Protagonism and Imperial Pleasures in Game Design #DIGRA21 | by meghna jayanth | Medium

Jump Over the Age (2022): *Citizen Sleeper*. Melbourne: Fellow Traveller.

Le Guin, U. (2014): 'Speech in Acceptance of the National Book Foundation Medal for Distinguished Contribution to American Letters'. Available at https://www.ursulakleguin.com/nbf-medal

Mad Max franchise (1979-present): Directed by George Miller.

Mainwaring, Ċ./Mulvey, G./Piacentini, T./Hales, S./Lamb, R. (2020): 'Migrant Solidarity Work in Times of 'Crisis': Glasgow and the Politics of Place', *Nordic Journal of Migration Research*, Special Issue: Vol 10, Issue 4.

McCarthy, C. (2006): *The Road*. New York City: Knopf.

McKendree, T. (1998): 'The logical core architecture'. *Nanotechnology*, Volume 9, Number 3

Metacritic (2025): *Citizen Sleeper* page. Available at https://www.metacritic.com/game/citizen-sleeper/

Rosenburg, U. (2007): *Agricola*

Sachs, J. (2018): 'We are all climate refugees now', *Project Syndicate*. Available at https://www.project-syndicate.org/commentary/climate-change-disaster-in-the-making-by-jeffrey-d-sachs-2018-08 (Accessed: 14th November 2024)

Simmons, D. (2003): *Ilium*.

Tsing, A. (2012): 'Unruly Edges: Mushrooms as Companion Species: For Donna Haraway'. *Environmental Humanities*, 1 (1).

Vella, D. (2010): 'Virtually in Ruins: The Imagery and Spaces of Ruin in Digital Games'. Masters Dissertation, University of Malta.

Vella, D./Cielecka, M. (2021): "You Won't Even Know Who You Are Any More: Bakhtinian Polyphony and the Challenge to the Ludic Subject in Disco Elysium." Baltic Screen Media Review, 9(1). pp. 90-104.

Vervoort, J. (2023): 'The mushroom at the end of the universe: feeling *Citizen Sleeper*'. Anticiplay project. Available at: https://anticiplay.medium.com/the-mushroom-at-the-end-of-the-universe-feeling-citizen-sleeper-8c3cd1aad24f (Accessed: 14th November 2024)

Wälivaara, J. (2018): 'Marginalized Bodies of Imagined Futurescapes: Ableism and Heteronormativity in Science Fiction'. *Culture Unbound*, 10(2), pp. 226–245. doi: 10.3384/cu.2000.1525.2018102226

ZA/UM (2019): *Disco Elysium*. Estonia: ZA/UM.

Ruin
A call to becoming-at-home again

Daniele Monaco

Abstract

This essay aims to combine Heidegger's conceptualisation of being-in-the-world and dwelling to the decay of cities and buildings – in sum, places. I argue that this decay can be examined from a phenomenological standpoint, distinguishing between ruins as mere debris and ruins capable of conveying an existential value to us. Rather than defining ruin by its function, state of decay, cause of destruction, or the antiquity of the structure, I propose to conceptualise it as a building that embodies the existential mood of ruinance (Ruinanz), concept proposed by Heidegger. By ruinance, I refer to the persistent existential awareness of the decay inherent in human existence and the world we inhabit. In this sense, ruins can be understood as decayed places that symbolise ruinance and evoke the "unhomely" (Unheimlichkeit). The existential value of ruins, as embodiments of ruinance, is connected to the primordial mode of existence that Heidegger identifies as "not-at-home" (Un-zuhause). Following Heidegger's proposal, I further contend that this uncanniness and homeless feeling, rather than inducing despair, can act as a call to action – to care for, dwell within, and ultimately attain a sense of "being-at-home" (Zuhause-sein) through engagement with ruins. This perspective challenges the conventional assumption that ruins are worthy only of neglecting, erasing, or abandoning them to further decay. To elucidate this distinction, I will explore two case studies of caring for ruins, both centred on community spaces. The first involves ethnographic research into the revival of a small-town centre in Ripa, Italy, while the second is an autoethnographic analysis of the video game Stardew Valley.

Keywords

Ruins, ruinance, community, dwelling, videogames world

1. Introduction

Many of the places we inhabit today no longer serve the purpose for which they were originally intended. These may include old, neglected structures in urban areas – such as train stations, public offices, and hospitals – that evoke feelings of unsettlement due to their deteriorating condition. This phenomenon is particularly prevalent in once-thriving urban centres, which have now become degraded spaces where inhabitants, if any, endure inhumane conditions (cf. Hollis 2013; Andersen 2019). In contrast, certain types of ruins, such as castles or palaces, continue to captivate us (cf. Tolia-Kelly/Waterton/Watson 2018). This raises several questions: why do these places elicit such different responses – unsettlement in the first case, allure in the second? And why do we deem one state of a building or place as inherently better or worse than another?

A potential response lies, according to Henderson (2010), in our belief that the built environment, especially in urban settings, must meet our desires, needs, and expectations. This belief is grounded in the assumption that a building's function – whether it provides shelter, healthcare, or other institutional services – defines its worth. If a building no longer serves its original purpose, we might label it a "ruin" without considering deeper implications. This perspective is influenced by our moral and economic frameworks, which are central to how we perceive and interact with these spaces (cf. Sennett 2018; Soja 2010). However, this view is limited. A building may still fulfil its intended function, albeit less efficiently, or it may even take on a new purpose. What truly disrupts our perception is not just a building's functional capacity, but its state of decay, which often evokes an uncanny and unsettling feeling.

Given these premises, I employ Heideggerian philosophy, spanning from *Being and Time* (Heidegger 1996a [1953])[1] to *Building, Dwelling and Thinking* (Heidegger 1993a [1951])[2], to understand the significance of ruins and how to live within them and taking care of them. I begin by exploring various perspectives on ruins and their significance, drawing on the works of Goethe (1970), Lee (1925), Simmel (1958), and Manca (2020). Then, I delve into Heidegger's interpretation of ruins, first outlining the core concepts of his philosophy and then focusing on his understanding of ruinance. Drawing from the concept of authenticity (Heidegger 1996a [1953]), I argue that ruins hold ontological significance, and that the act of

---

1   *Sein und Zeit* was first published in 1927. The text underwent several revisions and substantial modifications by Heidegger, with the final revision being the 12th edition of 1972. Since this essay does not aim to analyse the differences between editions, the English translation used here is based on the 7th edition of 1953, originally published by Niemeyer Verlag, which is very close to the 1972 version.
2   *Bauen, Wohnen, Denken* is the title of a lecture Heidegger delivered in 1951 in Darmstadt. Originally, the event was not intended specifically for specialists in philosophy but was aimed at engaging also architects and other practitioners.

taking care of ruins brings not only to a regeneration of a place, but also to live authentically (Heidegger 1996a [1953]: 240). Additionally, I propose a distinction between debris and ruins. Finally, I present two case studies – Ripa, a town in Italy, and the video game *Stardew Valley* (ConcernedApe 2016) to illustrate how the movement of care towards ruins can be tangible. In doing so, I draw a parallel between an example of caring about ruins in the real life and one in the virtual world, observing how the resulting conceptualisation of ruinance is applicable to both dimensions.

2. Different approaches to ruins

Fascination of ruins
Ruins became a central topic during the Romantic period of the 18th century. One of the main examples is comprised by Goethe's *Italian Journey* (Goethe 1970), completed around 1788, which highlights the allure ruins can evoke. Later, in the 1920s, Violet Page, writing under the pseudonym Vernon Lee, published a collection of essays on the *Genius Loci* (Lee 1925), illustrating how the captivation with ruins persisted in the 20th century. However, not all ruins inspired this kind of fascination (Augé 2006: 22) even still today ruins have an undeniable "seductive lure" (cf. Johnson 2014). One key to understanding this difference is to consider how Goethe and Page approached their journey to Italy. Italian ruins were not regarded merely as ancient buildings; rather, they symbolised a living past, deeply interwoven with contemporary life and architecture. Both authors observed that these ruins were often still inhabited and actively used, representing a coexistence of two distinct temporalities – one rooted in the present, with its living inhabitants, and the other brought to life in their imagination through the perception of the ruins. Consequently, we can identify two kinds of fascination with ruins: one that perceives them as a 'living past' and another that regards them as symbols. This interplay between the lived present and the evocative past gave the ruins a unique vitality, one that remains central to the ruins' philosophical and aesthetic significance (Goethe 1970; Lee 1925).

The temporal approach: the true time of Ruins
Goethe and Page's reflections make the conceptualisation of ruins more complex. In which time does a ruin become a ruin? Can a place be being seen as a ruin if it is still actively used? Most importantly, in which time a ruin truly belongs to? I believe that understanding ruins' temporality can be central to understand what a "Ruin" is. From a phenomenological standpoint, as presented by Edmund Husserl (cf. 2015 [1901]; 2011 [1893 -1917]), ruin belongs to the present while simultaneously existing as a trace or a document of the past: "on the one hand, it is no longer; on the other, the remains of the past hold it still present-at-hand *(Vorhanden)*. The paradox of the "no longer" and the "not yet" [...]" (Ricoeur 1988 [1983]: 77).

Moreover, the allure of ruins lies in their ability to evoke a living past, serving as a symbol[3]. Thus, a symbol operates on two distinct levels of reality simultaneously, allowing different aspects of reality – or, in this case, different times – to coexist. Martino Manca, in his essay *Rovine. Una tipologia e un tentativo di definizione* emphasises how a ruin could be analysed under the lens of their mystery of time. A ruin can bring into the present the perception of a past that therefore ceases to be past to became present:

The ruin thus bears witness to a new historical time which puts man before the change of history and in history and which is not limited to contemplation passive of the past as already dead, but on the contrary continually imposes to give meaning to what has lost meaning, but which cannot be concealed because there is in all its undeniable presence. (Manca 2020: 148)[4]

A series of questions arises when we try to understand the mystery of ruins from this perspective. First, there is the hermeneutic and phenomenological problem to experience a time in itself and not as past or memory, nor perceptual apprehension or remembrance and or recalling of the past.[5] It is instead, as stated by Manca, the meaning that we can give to the past summoned by the ruins. This is a problematic assumption since this rupture in the present time is filled by a meaning given to ruins through the relation between the subject the ruins. This calls into question intersubjectivity, as it requires a relationship between two entities[6]. We are unable to access the past inherent to ruins directly, and we experience it exclusively from our own present perspective. Nonetheless, the meaning of this past and or the meaning of the ruins differs for each of us, problematising the capacity of ruins to symbolize their original time (cf. Tolia-Kelly/Waterton/Watson 2018; Becker/Trigg 2025).

---

3   "Symbol" originates from the Greek words *syn* (together) and *ballein* (to throw), meaning "to throw together". ('Symbol', n.d.). In this context, I use this term to indicate the coexistence of two distinct levels of reality.

4   "La rovina testimonia quindi un nuovo tempo storico che pone l'uomo davanti al mutare *della* storia e *nella* storia e che non si limita alla contemplazione passiva del passato come già morto, ma anzi impone continuamente di dare un senso a ciò che un senso l'ha perso, ma che non si può celare poiché *c'è* in tutta la sua incontestabile presenza." Unless otherwise indicated, all translations are by the author.

5   According to Husserl, this involves the acts of *retention* and *protention* within our consciousness. (cf. Husserl 2011 [1893-1917]; Zahavi 2005).

6   In phenomenology, intersubjectivity is not merely a matter of mutual understanding but entails the capacity for two subjects to coexist in the same place and share a common perspective on the world. (cf. Husserl 2012 [1913]; Duranti 2010). In my view, this raises significant issues concerning ruins and their relationship with time and with us.

The Aesthetics approach: Ruins as struggle between nature and spirit

In his essay *Die Ruine,* George Simmel offered a theoretical perspective on the phenomenon of ruins. According to the author, a building symbolises our effort and capacity to create something – a "thing" – that did not previously exist. It is an expression of the poietic ability unique to human beings. According to Simmel, the decay of a building is not merely a form of wear and tear; it is an embodiment of the ongoing struggle between "nature" and "spirit"[7].

[...] our sense that these two world potencies the striving upward and the sinking downward-are working serenely together, as we envisage in their working a picture of purely natural existence. Expressing this peace for us, the ruin orders itself into the surrounding landscape without a break, [...] (Simmel 1958 [1911]: 384)

Any building is the product of various elements. "Nature" refers to the foundation – without it, nothing can be constructed. Its material aspect (stone, metal, wood, etc.) consists of elements that we cannot create *ex nihilo*. In other words, "Nature" is a movement, an activity distinct from human agency. "Spirit" is the poetic capacity of human beings to project, produce, and bring things into existence, which, in this case, shapes the material through architectural activity. With this theoretical framework, when Simmel speaks of ruins, the entity he analyses is a specific kind of ruin: one caused by the work of nature. Therefore, ruins become peaceful places where human beings no longer see themselves as violent imposers, enemies of nature (Simmel 1958 [1911]: 384). These themes were quite common during this time. Although not unlike reducing a building to its mere function, this approach reveals its limitations. In this case, the very essence of the building is understood through two pre-existing categories of reality: spirit and nature. Only through this distinction it becomes meaningful to interpret ruins as a truce, or compromise, between the two. The human being, allegedly belonging to the realm of spirit, but at same time material and natural, is at best a witness to this truce – while ruinance is inescapable, no architectural project is made to create a "ruin" – and at worst a violent aggressor against nature – as humans force it in a given form for their own needs. Bringing forth this perspective on ruins is no easier in contemporary times, as they have been commodified as capitalist destinations (Manca 2020: 148) within a broader perception of Western society in crisis – one that has also influenced thinkers as Heidegger (Ermarth 2000:

---

7 In his essay, Simmel employs the dialectic between *Natur* and *Geist*. The translation of *Geist* into English is debatable. It is sometimes rendered as "mind" or "soul", though these translations can be problematic due to their psychological connotations (Boy 2021: 191). In this essay, I use "spirit", as it more closely aligns with the philosophical categories implied and is consistent with the English translation cited in this work.

382). Ruins lose any sense of identity capable of conveying emotion or awakening our consciousness to their existential value. The way in which Goethe, Simmel or Page felt Ruins, the *Stimmung*[8] of that time, does not exist anymore.

2.1 The ontological approach: the theoretical framework

In this article, I propose that it is possible to grasp the existential meaning of ruins by engaging from an ontological viewpoint. Among other philosophers, Martin Heidegger has analysed several key concepts that can explain the ontological status of ruins. The following paragraphs analyse core conceptualisation of his philosophy that are useful for my analysis.

Life as fallenness

According to Heidegger, our way of existing is shaped by our unique ontological nature (Heidegger 1996a [1953]: 49). This nature is so distinct that Heidegger avoids even the term "human being" instead using the word *Dasein* (ibid: 15). *Dasein* literally means "being-there" but "there" here refers not to a specific place, but to the "world" as a foundational ground – an *a priori* condition necessary for the very existence of human being. Thus, *Dasein* also implies "being-in-the-world" (*in-der-Welt-sein*) (ibid: 49). This means that every encounter with intramundane beings take place not inside a world as a box with items placed casually[9], but instead through our being in the world[10]. Now, "being-there" and "being-in-the-world" are so close to be almost the same thing: there is not such a thing as a *Dasein* that it is not immediately in and with the world, and a being which nature is to being-in-

---

8 Typically translated as "Mood" *Stimmung* refers to the existential mode through which we understand and feel the world from within, prior to any conceptualisation or emotion. (Dahlstrom 2013: 133)
9 Understanding the world as a physical extension containing objects has been a part of philosophical thought at least since Descartes. He referred to the physical and material extension of objects, and physical space as extension, as *res extensa*. (cf. Descartes 2013 [1641]).
10 For the sake of this essay, I intend to define "World" as "[...] set composed of beings that are understood together with all their properties and mutual relationships. [...] the horizon (or ground) against which every object is experienced and understood" (Gualeni/Vella 2020: 27).

the-world that it is not a *Dasein*[11]. While non-human beings, such as animals, exist *within* the world, only *Dasein* is *with* the world and must actively engage with it[12].

Dasein engages with the world and discloses it, according to Heidegger, through the fundamental state of anxiety (*Angst*). He identifies *Angst*, unease, anxiety, malaise (Dahlstrom 2013: 15) not merely as a perception or mental state but as a mood (*Stimmung*) that is in this case intended as fundamental and primary way of existing and relating to the world. *Angst* reveals the world as world in its entirety. Moreover, it is only through this ontological state that we can perceive the world as it is in itself, rather than merely as a collection of objects. The fundamental mood of *Angst* allows us to confront and accept our "thrownness" (*Geworfenheit*) (Heidegger 1996a [1953]: 164). Thrownness implies that, as *Dasein*, we do not choose to be born or to exist, nor can we choose not to exist. We cannot determine the place and time of our existence or the ground upon which we stand, the range of our fundamental possibilities during our lifetime[13] and that we exist as mortals, and our "being-toward-death" (*Sein-zum-Tode*) (ibid: 219) is beyond our choosing or undoing. In other words, we exist in the world without having chosen it, and through *Angst*, we become aware of the world itself and our ontological condition. This ontological state signifies that, as *Dasein*, we exist solely within and in relation to the world into which we have been thrown. Our fundamental mood, *Angst*, serves both as a means of revealing the world and as a call that awakens the human being to their existential condition (ibid: 247). But why is such a traumatic and unsettling experience necessary to awaken *Dasein* to the reality of the world? According to Heidegger, this is because our thrownness also predisposes us to "fallenness" (*Verfallenheit*): "in the self-certainty and decisiveness of the they, it gets spread abroad increasingly that there is no need of authentic, attuned understanding. The supposition of the they that one is leading and sustaining a full and genuine 'life' brings a tranquillization to Da-sein, for which everything is in 'the best order' and for whom all doors are open" (ibid: 166). In other words, we constantly risk becoming absorbed, seduced, or alienated by the things within the world – either by presuming, without question, that we are fully *at home* in the world, or by resigning ourselves to the belief that we can never truly find a place of belonging within it. The interesting thing about it is that Heidegger in his early work calls this movement of life against itself ruinance.

---

[11] The equation between the human being and *Dasein* is robust but not absolute. Scholars hold differing positions, with some equating the two and others drawing distinctions. In this essay, I assume that *Dasein* and being-in-the-world, while not necessarily identical, are at least inseparable.

[12] This is a well-known thesis from Heidegger: human beings are "world-forming", animals are "poor in world" and objects are "worldless". (cf. Heidegger 1995 [1930]).

[13] Heidegger does not mean that we cannot plan our lives or make decisions that shape our unique biographies. Rather, he argues that all these aspects of our lives are grounded in our ontological nature—one that we cannot choose or alter.

3. Ruins as symbol of decaying of life: proposal on Heideggerian philosophy

In this paragraph, I aim to demonstrate how connecting various concepts from Heidegger's theory – each of which has been individually analysed by scholars as Ruin (2012), Manca (2020), Murchadha (2002) yield a surprisingly coherent and effective ontological perspective on ruins. While scholars have extensively explored these concepts in isolation, they have yet to synthesize them into a unified framework that directly addresses ruins. This synthesis becomes possible through the introduction of the concept of ruinance from Heidegger's early work, alongside the ideas discussed above. Moreover, after engaging with the definition of ruins, I argue that it is also possible to apply the ontological nature of ruins to the world of video games.

A different nuance of our life: Ruinance as fundamental state

During his 1920/1921 course titled "Phenomenological Interpretation of Aristotle", Heidegger introduced the term "ruinance" (*Ruinanz*) (Heidegger 2009 [1921]). This term appears only briefly in his work, and as far as is known, it disappears from his later writings (Ruin 2012: 16). According to Heidegger, we become aware of the nature of our existence when we realise that life "propagates as an on-going collapse, plunge, or fall" (Ruin 2012: 18). According to Heidegger (and Ruin's interpretation), ruinance reflects an existential trait of life itself. Our existence is limited, mortal, and finite, bound to both the world and to its inherent impermanence (Ruin 2012: 19). Life here encompasses our entire existence, including its projects, decisions, activities, and, crucially, the things brought into appearance, such as buildings[14]. Ruinance can be described therefore as the decay of the movement of life itself, which include all the things life can do within and in the world. Furthermore, ruinance, much like thrownness, can give rise to a fundamental state of uncanniness – a profound anxiety that can serve as a call to something beyond itself.

Heidegger describes the state of *Unheimlichkeit* as a nuance within the spectrum of fundamental states (Heidegger 1996b [1942]: 144; 1930). In this context, *Heimlich* relates to what is familiar and safe, deriving from *Heim*, meaning "home" – a place to live. However, *Heimlich* here loses its spatial link, and so *Unheimlich* takes on the meaning of "uncanny" though it literally means "unhomely". The dual nuances of this fundamental ontological state demonstrate the complexity of our language in describing it. "Home" is an existential category, therefore not related to the materiality of the world, representing what is familiar,

---

14 The question of technology, how we can produce things, in Heidegger is a complex topic. (cf. Heidegger 1993b [1954]; Plunkett 2021).

safe, and comforting, our right place in the world. Therefore, *Unheimlichkeit* signifies a sense of "not-being-at-home" and the unsettling sting of the uncanny that arises from feeling out of place in the world[15]. In the later Heidegger's work, the concept of not-being-at-home is still present but with a different nuance (Kurir 2022) more specifically related to the capacity of human being to dwell the world. In the next paragraph, I intend to argue how a building can have an existential meaning and, therefore, how can a ruin become the physical aspect of ruinance and give us the fundamental state of unhomely.

The connection between ruinance and buildings

In 1951, Heidegger delivered a course titled *Building Dwelling Thinking*[16], where he challenged the assumption that building is merely a technique act of constructing structures for shelter. For Heidegger, building is an essential part of how we inhabit and relate to the world. Heidegger posits that to build in a point in the space (*Stellen*) means to create a place (*Ort*) that can create a new kind of space, a space (*Raum*) that can actually "make space" for a place that foster true dwelling (Heidegger 1993a [1951]: 354–55). Dwelling, in this sense, is not about physical occupation of space but about engaging in a meaningful relationship with the true existential meaning of the building. In this way, Heidegger redefines building as an ontological activity, a practice that shapes spaces that allow us to dwell fully, to live authentically, and to experience the world in its essence. In this sense, a building is always an ontological relationship with the world. Therefore, it is possible to build a home in authentic relationship with our world (ibid: 361–62). A place, a building, is always something that allow us to engage with our ontological situation. To dwell, for Heidegger means building, cultivating and take care of. All this path of activity does not show a precise hierarchy or timeline: as human being, we always dwelling, taking care and building (ibid: 349).

It is here that we can trace a movement between ruinance, ruins, and not-being-at-home. Building allows us to become at home – to be familiar with, understand, and take care of our world. In other words, to be at home. The "home," as building and ontological category, becomes the centre of our place in the world; we make ourselves at home in a world we learn to dwell within (Norberg-Schulz 1980: 170). The primary condition of *Dasein* is to not-being-at-home, therefore I propose assigning a different significance to the movement between not-being-at-home and becoming-at-home, connecting this movement to buildings. This process is, almost literally, a movement, as Caterina Resta describes it:

---

15 (Heidegger 1983 [1953]: 116–27). Cf. Dahlstrom (2013: 228).
16 *Building, Dwelling, Thinking* in (Heidegger 1993a [1951]).

> *Being-of-home (Heimischsein) is therefore not the starting point, nor should the endpoint be understood as mere being-at-home (zu Hause); rather, it is always a matter of becoming-of-home (Heimischwerden)* (Resta 2020: 127)[17]

In other words, *Dasein* always begins as thrown into the world and, through the ontological activity of building, can become at home. However, in the connection between the ontological movement of life and the existential significance of building, ruins are notably absent. This absence highlights a gap in Heidegger's analysis: an ontological approach to buildings that are no longer in their original state. It is here that the concept of ruinance can be employed to meaningfully engage with ruins.

3.1 The ontological definition of ruins

According to Heidegger, constructing a building is an ontological act that allows the *Dasein* to become "at home". Being-at-home (*Zuhause-sein*) begins from a fundamental state of not-being-at-home. While the existential state of being-at-home finds its physical correlate in a well-constructed (as to say, authentically) building (Heidegger 1993a [1951]: 361), the role of ruins has often been overlooked. This is where the concept of ruinance becomes crucial. Ruinance, as the experience of limitation, decay, and finitude, reflects our way of existing in the world. Life is inherently marked by finitude, and a ruin manifests this existential truth. A ruin is the external manifestation of this *ruinance*. If life – understood as our ontological way of existing in and with the world – discloses itself as *ruinance*, with its inherent impermanence, then ruins are no longer merely states of decay, defined in opposition to some prior condition of wholeness. A ruin, by contrast, represents a moment – the counter-movement – within the ontological unfolding of the poietic capacity to build; that is, to exist and to create a home in the world.

As with all aspects of human existence, a building, like *Dasein*, is governed by nothing other than its own ontological nature. A newly constructed building and a ruin are two expressions of the same existential connection between human being and buildings. Hence, I argue that the decay of a building – whether due to time, neglect, or natural disasters – is not a failure of human projects or a struggle between humanity and nature (Simmel 1958 [1911]). Rather, it reflects the inherent finitude of all beings within the world. Decay, in this view, represents the *ruinance* intrinsic to life's very movement. Therefore, it is possible to connect the missing link within the process of becoming-at-home. If building constitutes the "positive" moment of this movement – allowing *Dasein* to be at home – the

---

17 "L'essere-di-casa (*Heimischsein*) non è dunque il punto di partenza; e neppure il punto d'arrivo è da intendersi come un mero essere-a-casa (*zu Hause*); si tratta piuttosto sempre di un *divenire*-di-casa (Heimisch*werden*)".

"negative" moment, when the human being is not at home or has lost their home, remains unexplored in Heidegger's interpretation (cf. Heidegger 1993a [1951]). I argue that just as a building allows us to be at home, ruins can serve as the expression of *Unheimlichkeit* – the uncanniness of not-being-at-home and, therefore, as call to take care of ruins.

Ruins, when genuinely experienced as the physical manifestation of life's movement, can act as the voice of *Unheimlichkeit*. They remind us of the existential need to continually becoming-at-home and explain why, in our everyday lives, encountering buildings in a state of decay can attract our attention and evoke discomfort: they serve as the voice of this existential mood. This understanding repositions ruins, not merely as decayed remnants but as integral to the dynamic interplay between being-at-home and not-being-at-home in human existence. Thus, the existential trait of care (*Sorge*) (Heidegger 1996a [1953]: 178; 279) is called upon to engage with ruins authentically, assigning them meaning within our relationship to the world rather than relegating them to something broken or undesirable. Taking care of ruins, therefore, becomes synonymous with taking care of our very existence as human beings in the world. In other words, taking care of ruins can means choose, as *Dasein*, to exist authentically[18].

Such conceptualisation of ruins can take various forms, but I believe a distinction is necessary from a phenomenological viewpoint. A "Ruin", in the proper sense, is a building with which human beings can form an ontological connection. It is a structure in which we recognize, even if it consists merely of scattered bricks, crumbled stones, or fragments, the expression of our existence and the inherent movement of life as ruinance. Therefore, what allows us to give a meaning to ruins, to hear their voices, it has always been laying in the relationship with them. "Debris", instead, would be a building which this ontological relationship cannot be engaged and therefore this type of decayed building will not call human being to the "duty" to take care of the existence of this thing. Within this theoretical framework, even a brand-new building could be considered debris, whereas a decayed home, if lived in and cared for, could be a ruin. In other words, it is essential to emphasise that the relationship between ruins and debris pertains specifically to buildings in a state of decay and how they can easily transform into one another. There is no sharp distinction between a building as debris and the

---

18 Both Manca and Ruin focused on the idea that ruinance can be used to explain the way in which human being exists in the world, and even how, according to Manca, our interest for the ruins can suggest some connection with the fundamental state of ruinance. On the other hand, Felix Murchada, in *Being as Ruination: Heidegger, Simmel, and the Phenomenology of Ruins* (2002), explores the possibility of understanding ruins through Heidegger's philosophy. However, he chooses to focus on the existential meaning of buildings rather than engaging with the existential category of ruinance.

same building as a ruin or home. Rather than posing a problem, this transformative capacity of the ontological relationship is precisely what enables us to consider the potential for caring for ruins without relying on deterministic concepts such as "intended function", appearance, comparison with an original design, or a nostalgic infatuation. Only through this ontological movement can we engage with the motility of being-in-the-world, and it is here that the process of becoming-at-home is grounded. Without this, ruins could only be erased or forgotten.

4. The case studies: the call of the ruins

The aim of this essay is to conduct an unusual experiment in philosophical inquiry by applying a theoretical framework on ruins to specific case study. Here, I examine a real-world example: Ripa, a small village in Umbria, Italy, where I lived for a year. During this time, I conducted semi-structured interviews and examined sources available in the Biblioarna library (Ripa's library). The goal of my analysis is to explore the existential implications of ruins. Moreover, my theoretical framework suggests that an ontological understanding of ruins is not confined to materiality in the "real" world but also applies to the fictional, or "non-existent" world. Therefore, in addition to Ripa, I also analyse a case study from a virtual world, the videogame *Stardew Valley*, published by ConcernedApe in 2016. Here, I focus on the community centre of Pelican Town, where the player begins their journey. I use an autoethnographic method to examine this virtual space, drawing from my own gameplay experience.

4.1 Ripa: a community centre by choice

Ripa is a small village comprising approximately 1,000 dwellings, distinguished by its circular castle, which was constructed in 1266 during a conflict between Perugia and the nearby AssisiF[19]. Ripa's history is rich with intriguing events and notable figures across the centuries, beginning with the conflict between Perugia and Assisi (around 1202), followed by the actions of famous bandits and mercenaries around 1400 (cf. Tufo 2008), as well as ongoing struggles between the authority of Perugia and the Pope. In 1280, in the middle of the castle, the Church of S. Andrea was built, and beside it, the aedicula of S. Emiliano. These two buildings served as the headquarters for two distinct dioceses, each governing

---

19 Giuseppe Tufo, scholar and dweller of Ripa, is currently investigating a possible connection with the Order of the Knights Templar. Through a thorough examination of historical documents, he is exploring links between the unusual shape of the fortified settlement and the discovery of alleged Templar symbols around historical buildings.

a theoretical half of the village. This unique arrangement (cf. Tufo/Bastianelli 1998) has inspired two *contradas* (neighbourhood teams) that nowadays engage in friendly competition during medieval games held at the village festival, the *Palio di Ripa*.

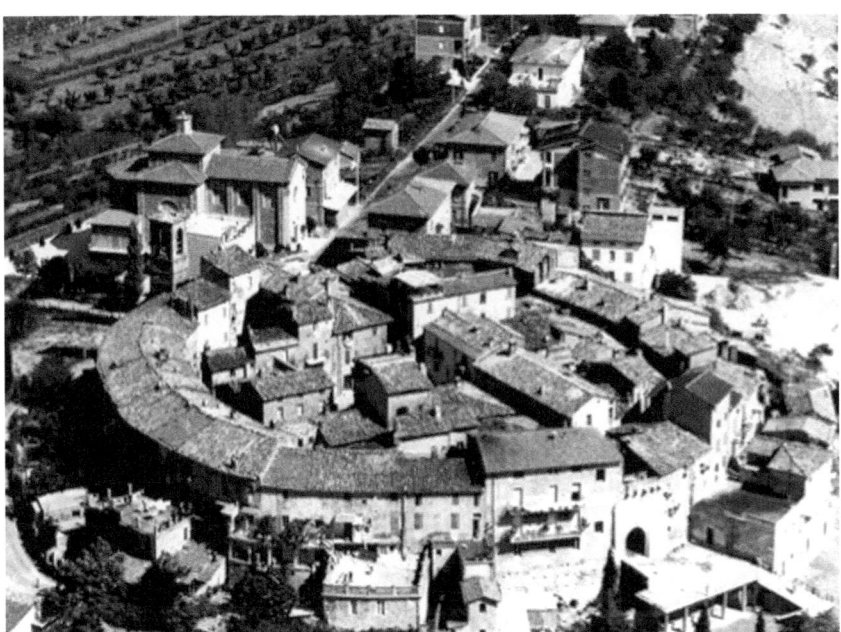

Fig. 1: Aerial picture of Ripa. 1970. Courtesy of Associazione Federici, ProArna, Biblioteca Federici.

By the late 19th century, the premises had fallen into disrepair and were sold to the municipality of Perugia. The diplomat Giovan Battista Pioda restored the remain original structures and donated them back to the community and the community established the Palmira Federici Association (cf. Bastianelli 2008).

The second floor housed the municipal registration office, and in the 1960s, it briefly functioned as a kindergarten. In 2010 the building was destinated to be a library named *Biblioarna*, still active today (cf. Bastianelli 2008).

I had the chance to live in Ripa for almost a year, between January and October 2023, and to be in contact with its inhabitants. What started as a practicality – going to the Biblioarna library to proceed with my studies – soon became a source of curiosity and inspiration. During the past years, I started to collect memories with the aim of building an autoethnography (cf. Adams/Ellis 2017) of the Ripa community to integrate it in my PhD project, from which also the following reflections stem. Between August 2023 and October 2023, I interviewed formally 6 people involved directly in the activities of library Federici and spoke with countless dwellers of Ripa.

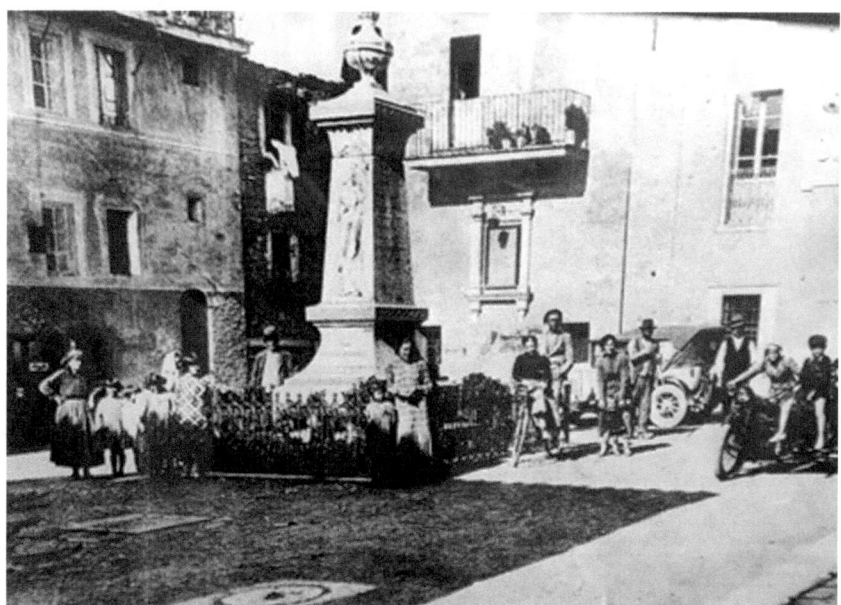

Fig. 2: Ripa community centre, 1934. Probably Afterwork Association. Courtesy of Associazione Federici, ProArna, Biblioteca Federici.

Ruins in Ripa

The case of the "community centre" of Ripa is particularly relevant because this building has undergone all the possible approaches discussed above for understanding its value as a ruin. Over time, the building has repeatedly changed its function, often transitioning between activities so disparate that it is difficult to claim that the inhabitants of Ripa have engaged with it in a consistent manner or with the same sense of purpose. For example, the children who attended the building when it served as a school likely had experiences shaped by formative learning, bonding with classmates, and daily routines. These experiences would differ significantly from those of individuals who worked there when it was repurposed as an administrative headquarters – perhaps dealing with the frustrations and monotony of bureaucratic tasks. Interestingly, even the "romantic" interpretation of the building as a ruin does not seem to resonate with the community. In fact, in a historically Catholic context like Ripa's, one might expect the fact that the two buildings were once sacred places to imbue them with a romantic aura, fostering a sense of collective memory and community identity. However, this is not the case. The buildings changing functions and appearances over time have rendered such associations obsolete. None of the individuals I interviewed referred to their former sacred nature as a source of sentimental or symbolic attachment; it was mentioned, at best, as a point of historical curiosity.

While the physical restoration of the building itself was undertaken by external entities, such as the municipality of Perugia, the psychological and communal investment reached its height in the last two years. During this time, many residents dedicated significant amounts of their time and energy – on a voluntary basis – to sustaining the activities of the centre. What emerges from the interviews is that, at some point, the residents found themselves in a position where they risked losing their sense of home. According to them, Ripa was effectively abandoned by the municipality. It is a fact that, over time, the municipality of Perugia progressively shut down the local branch of a bank and attempted to do the same with the local office of the Italian Postal Services. As is often the case in small towns and villages, such services serve as crucial support points, particularly for the elderly, who rely on them to manage tasks such as banking, paying bills, and other essential activities. Moreover, some of its inhabitants used only to live in Ripa because, for its position, it is cheaper then live in Perugia. Still today the only relationship between them and Ripa is one like where to sleep or rent a cheaper house. Therefore, some of the dwellers, especially who use to live inside the castle, feeling that they should rely only on their own, and looked around and saw the library as opportunity. Here the theoretical framework proposed provides a hermeneutical key to understanding why the library became the centre of a new effort.

The building was initially a debris, a space without purpose, identity, or any capacity to convey a sense of belonging. In my opinion, the change that mattered was not the different functions the building has served but rather how the community has come to perceive it over the past two years. The building came to be seen as a ruin in the sense that the community recognised it as a symbol of the decay of Ripa, the physical embodiment of the area's decline. According to Heidegger, it is only when *Dasein* recognises and embraces its own nature that it is possible to live authentically (Heidegger 1996a [1953]: 282; 2009 [1921]; Ruin 2012: 28). In this case, I think that when the community recognised the true nature of the building as ruin, they also began to take care of it. In doing so, they were, in turn, taking care of their own way of existing in the world, since, due to the existential meaning of buildings, taking care of the building meant taking care of their relationship with the world. The Biblioteca Federici spoke as ruin and the community listened and this relationship can show the ontological ground of all the effort of the community.

Today, efforts are underway to involve a new generation of *ripaioli* (residents of Ripa), focusing on enhancing the role of the Biblioteca Federici and others meaningful places of Ripa in bringing the community together. These initiatives aim to strengthen local identity and emphasise the importance of preserving the heritage of the place.

4.1 Dasein in Virtual World

Before introducing the next case study, it is necessary to explain how this theoretical framework can be found inside virtual world. It is problematic, from a philosophical viewpoint, to argue that this complex ontological movement of life, and the ontological meaning of ruins, is still stay in place in a virtual world such as videogame world. Although, the idea that game worlds themselves could be a possible subject of philosophical study is not new. Eugen Fink, for instance, dedicated much of his work to the notion of play, recognising its ability to create fictional worlds with an ontological presence, where individuals can explore beyond their original capabilities (cf. Fink 1968 [1957]). Indeed, he defined play as a "fundamental phenomenon" (*Grundphänomen*) (cf. Fink 1979), arguing that human beings are, by essence, players. The step from fictional worlds in general to the game world is a short one, and video games are now an object of philosophy of games (cf. Gualeni 2015; 2016). Now, one might have the impression that digital worlds are only endless and indifferent simulations, thereby becoming "non-places," both in the sense of lacking relational meaning and significance, and in the sense of escaping the "real" order grounded in the earth. However, the opposite attempt is also possible: one can grasp the existential importance game worlds hold (cf. Gualeni/Vella 2020; Gualeni/Fassone 2023). The field of games and digital studies has investigated the status of the virtual world, and among various possible interpretations, it offers the working definition,: a virtual world is to be understood as "[...] the (relatively) perceptually stable interactive experience that is disclosed by a computer-generated environment" (Gualeni/Vella 2020: 27) Hence, a virtual world refers to a computer-generated environment with which one can interact and to which we have continuous access.

Bring in the virtual world our ontological nature

Research on game worlds has shown how our activities thoughts, and subjectivity are maintained even when entering a game world. According to Vella, the player finds themselves experiencing games worlds as "ludic subject" (cf. Vella 2015). The player's engagement with the game world through a phenomenological embodiment in the form of the playable figure reflects the experiential and existential structures of the embodied being-in-the-world (ibid: 16). In other words, the notion of *Dasein* as being-in-the-world persists in the game world, and that, consequently, the human being as ludic subjectivity finds itself "thrown" as "being-in-the-game-world" where being "actual" means existing in the world from which I speak. (Ryan 1991: 18) Vella points out that the risk of such an approach to virtual worlds lies in misunderstanding the experience within videogames as merely an aesthetic experience. Instead, he argues, we should engage with virtual worlds through an ontological and phenomenological approach. In this context, dwelling in the virtual world continues to signify the "bringing-to-presentation" of existential practices that constitute the player's being-in-the-game-world (Vella

2016: 82) Building continues to be "in phenomenological terms, [...] the setting-in-stone of a revealed mode of being" (Vella 2019: 151). If, as Vella has argued in *There is No Place Like Home*, it is possible to incorporate ontological activity within Heideggerian philosophy into virtual worlds, then it is also necessary to address how existential traits of human beings – such as ruinance and the ontological significance of ruins – persist in these contexts[20]. In other words, the "there" of *being-there* must encompass virtual worlds in their entirety, along with all their existential dimensions. Following the scholarship on the philosophical relevance of virtual world this case study concerns a virtual place, a specific building in Stardew valley (ConcernedApe 2016)

4.2 Stardew Valley: a "natural" ruin in an "artificial" world

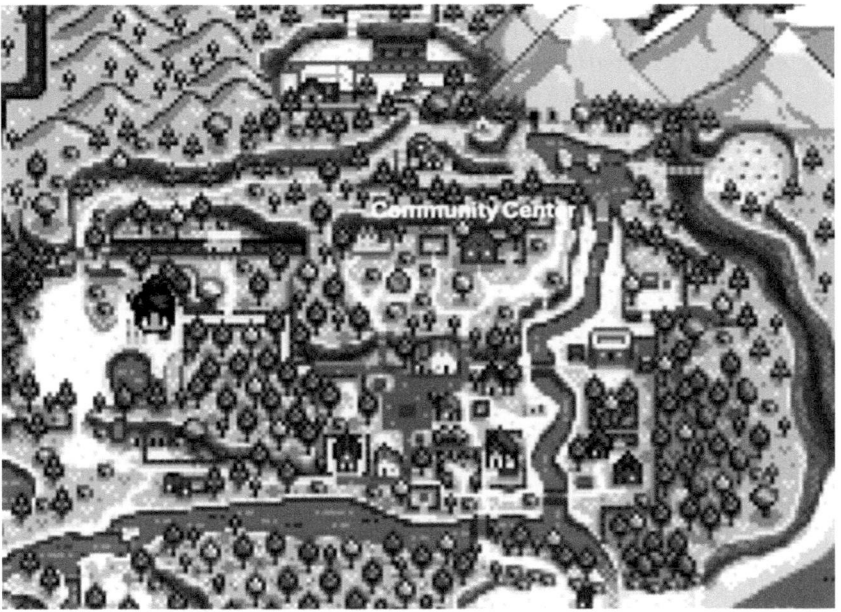

Fig. 3: The map of Pelican Town (Detail). Screenshot by the author.

The game was published in 2016 and is an open-ended country-life simulator, with a nonlinear gameplay and it is well-known among scholars (cf. Mackay/Roberts 2023). The opening scenes of the game depict the player working in a corporate environment that closely resembles the sterile, monotonous offices of

---

20   It is relevant to note that game scholars have started analysing the relevance of ruins and decay in game worlds, or even the value of games as ruin themselves. However, the relationship between philosophical conceptualisations and ruins has still been seldom analysed. (Fuchs 2017; Vella 2010)

well-known brands. The colour palette is dominated by greys and blues, with workers confined to cubicles. One cubicle even contains a skeleton, emphasising the lifelessness of the setting. The player character, exhausted and frustrated by this lifestyle, remembers a letter from their grandfather. This is how they decide to quit their job and move to the old cottage.

The player is free to engage in a variety of activities in whatever order they prefer. These activities include farming, fishing, foraging, mining, fighting monsters, crafting, cooking, gift-giving, completing quests, donating to the local museum, and restoring the community centre. The player also participates in key community events, such as the Egg Festival in spring or the Dance of the Moonlight Jellies in summer. These events, along with the changing seasons, structure the player's activities: each season and event make available unique resources that can only be collected during specific times.

The community centre in Pelican Town is presented as a central element from the outset of its narrative. Upon the player's first visit, they are guided by Lewis, the mayor of Pelican Town, ostensibly to investigate a "rat problem". Here, they encounter the Junimos – small, green forest spirits who agree to assist in restoring the centre. From this moment, the player is tasked with gathering a variety of items, such as crops, fruits, ores, and other resources, to complete specific bundles. Each completed bundle prompts the Junimos to reward the player by restoring a portion of the building. This restoration process requires significant time, often spanning at least two in-game years, due to the seasonal availability of certain resources and the effort involved in obtaining them.

The meaning of ruin of Stardew valley

The community centre demonstrates to the player that, despite the calming, almost cozy depiction of life in Pelican Town, parts of the town can be seen as a symbol of decay and, therefore, as ruinance. According to Heidegger, the *Dasein* is often absorbed in everyday life and idle talk (*Gerede*), avoiding confrontation with the ruinance inherent in existence (Heidegger 1996a [1953]: 157). This dynamic is mirrored in the world of *Stardew Valley*, where each NPC (non-playable character) appears immersed in a life that, while not perfect, is seemingly peaceful – a reflection of *Dasein*'s tendency to evade the recognition of life's transience and decay.

The town, however, faces a looming threat from a capitalist force embodied by the Joja Market, analogue to supermarkets in contemporary Western society. Pierre, the owner of the local shop selling seeds and tools, struggles to compete with Joja's low prices. Although his products are, allegedly, of superior quality, cutscenes reveal how even the residents of Pelican Town occasionally forsake local interests to shop at Joja Market (cf. Crowley 2023). This lack of collective care is symptomatic of the town's broader disengagement, a phenomenon not explicitly addressed by the community until the mayor informs the player about the state of the community centre. While the mayor does not directly accuse the townspeople of abandoning the centre, it becomes evident that its state of decay is a result of time and neglect. In this sense, the community centre represents an example of a ruin, even more so than the building in Ripa. Nevertheless, I argue that the same ontological framework can be applied to make sense of it.

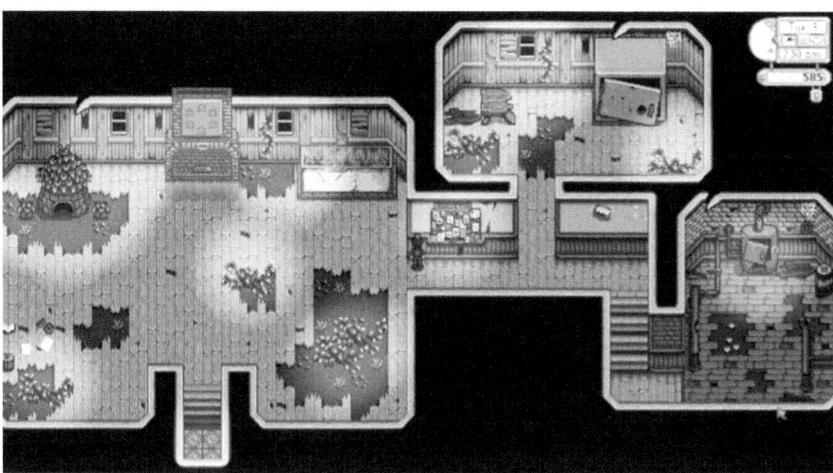

Fig. 4: Community centre, inside. Screenshot by Author.

When the player first encounters the community centre, the building is only a debris, devoid of any connection or existential meaning for either the community or the player. Upon completing its restoration, the player is faced with a significant choice: to fully restore the building and return it to the community, or, if they are a Joja Market member, to repurpose it as a storehouse. In the first scenario, the NPCs begin using the community centre again, and the Junimos hold a heartfelt farewell ceremony before departing. In the second scenario, however, the Junimos silently vanish, and the building loses its communal purpose, becoming a purely utilitarian space. Beyond the overt moral undertones implied by the game design, the most compelling aspect here is the player's pivotal role in this decision-making process. I argued, following Vella and Gualeni (2020), that the player, as *Dasein* or being-in-the-game-world, engages with the game world within all their existential traits. Thus, when the player perceives the community centre as a ruin, the enchantment of Pelican Town's idyllic life is disrupted, and the player experiences a sense of not-being-at-home. The uncanniness of a life absorbed by everyday routines is accompanied by an awareness of the ruinance of life, embodied by the decaying community centre. The game encourages the player to take care of the world of Pelican Town, with quests and side quests designed to have positive effects for both the player and the community. The player is integrated into a supportive network of friendships and relationships – if they choose to engage with the villagers. Here, the law of becoming-at-home is evident. The player learns their right to feel at home, not only through building, cultivating, and taking care – activities closely aligned with Heidegger's *Building, Dwelling, Thinking* – but also by tending to the ruin of the community centre. While the case study of Ripa required the community to repurpose the building with a new function, here the task is more traditional: the building needs to be restored to its previous form and function. This restoration process, however, mirrors the same existential movement of care and belonging, reinforcing the connection between taking care of ruins and establishing a sense of home. The voice of the ruin in this case it is not or mainly the community, but rather the Junimos. The player become agents within the law of becoming-at-home and, as *Dasein*, can choose either to engage with and embrace this ontological relationship or to avoid it, thereby living an inauthentic life – inauthentic in the context of the story's framework – in Pelican Town.

Conclusion

This essay analyses the Heideggerian understanding of ruins and proposes a philosophical framework on the existential interplay between being-at-home and not-being-at-home. Central to this argument is the concept of ruinance, the ontological condition of impermanence and decay that permeates human existence and finds its physical expression in ruins. By bringing together Heidegger's early phenomenological insights with the specific existential of ruins, aim at demonstrating how ruins are not merely remnants of decay (Augé 2006) or a truce between nature and spirit (Simmel 1958 [1911]) but are integral to the ontological movement of life.

Drawing from Heidegger's philosophy (1996a [1953]; 1996b [1942]; 1993a [1951]), constructing and dwelling in a building is presented as an ontological act that enables *Dasein* to become at home in the world. Yet, this process is not linear or absolute. It begins in a state of not-being-at-home and involves a constant negotiation between finitude and belonging. Ruins, as argued here, embody this negotiation in their very existence. They serve as the physical manifestation of ruinance, reminding us of the impermanence of life and the transient nature of all human projects.

The case studies – Ripa's and Pelican Town's community centre – illustrate how this framework can be applied to both real and virtual worlds, offering a deeper understanding of the ontological significance of ruins. In the case of Ripa, the community centre initially represented a state of neglect and loss, a debris devoid of purpose or identity. Over time, the building became a ruin in the Heideggerian sense, as the community began to perceive it as a symbol of their own decline and alienation. Recognising this allowed the residents to take authentic care of the building, repurposing it into a library and a hub for communal activities. This process of taking care was not merely an act of physical restoration but an existential act, as the community redefined their relationship with the building and, by extension, with their own sense of place and belonging.

Similarly, the community centre in *Stardew Valley* (ConcernedApe 2016) offers a symbolic narrative of ruinance. Despite its idyllic setting, Pelican Town contains a decaying building that embodies the uncanniness of not-being-at-home (Heidegger 1996a [1953]; 1996b [1942]), disrupting the apparent harmony of the town. The player, as *Dasein*, can choose to engage with the building and undertake its restoration. This process requires effort, time, and a deliberate acknowledgment of the decay, mirroring the existential care needed to reconcile with one's own impermanence. By restoring the community centre, the player not only revitalises the building but also strengthens their connection to the town and its inhabitants, embodying the law of becoming-at-home. This opens new parallels between the act of caring about the building and moving forward in the game, further reflecting on multiple ways in which one engage with dwelling in game space.

These case studies illustrate the universality of ruins and ruinance and its capacity to bridge the real and the virtual. Whether in a small Italian village or a pixelated farming town, the ontological dynamics of taking care, rebuilding, and engaging with decay remain consistent. They reaffirm that ruins are not static objects but active participants in the existential movement of life. By embracing this perspective, we gain not only a deeper appreciation of ruins but also a renewed understanding of our own finite, transient existence.

References

Adams, E/Ellis, C/Holman, S. (2017): "Autoethnography". In Matthes, C./Potter R. (eds.) The International Encyclopaedia of Communication Research Methods, pp. 1–11. https://doi.org/10.1002/9781118901731.iecrm0011.

Andersen, H. S. (2019): Urban Sores: On the Interaction between Segregation, Urban Decay and Deprived Neighbourhoods. 1st ed. Routledge. https://doi.org/10.4324/9781315191980.

Augé, M. (2006): Rovine e macerie: il senso del tempo. Torino: Bollati Boringhieri.

Bastianelli, G. (2008): Associazione "Palmira Federici". Cento Anni al Servizio Di Ripa. Ripa.

Becker, T./Trigg, D. (eds.) (2025): The Routledge Handbook of Nostalgia. London: Taylor & Francis.

Boy, J. (2020): "The metropolis and the life of spirit by Georg Simmel: A new translation". Journal of Classical Sociology. 21. pp. 188-202 1468795X2098063. 10.1177/1468795X20980638.

ConcernedApe. (2016): Stardew Valley. PC. ConcernedApe.

Crowley, S. (2023): "Playing Farmer: At the Intersections of Neo-Liberal Capitalism and Ecocriticism in Stardew Valley". Journal of Gaming & Virtual Worlds 15 (1), pp. 21–37. https://doi.org/10.1386/jgvw_00069_1.

Dahlstrom, O. (2013): The Heidegger Dictionary. Bloomsbury Philosophy Dictionaries. London: Bloomsbury Publishing.

Descartes, R. 2013 [1641]: Meditations on First Philosophy: With Selections from the Objections and Replies ; a Latin-English Edition. Cambridge: Cambridge University Pr.

Duranti, A. (2010): "Husserl, Intersubjectivity and Anthropology". Anthropological Theory 10 (1–2), pp. 16–35. https://doi.org/10.1177/1463499610370517.

Ermarth, M. (2000): "Heidegger on Americanism: Ruinanz and the End of Modernity". Modernism/Modernity 7 (3), pp. 379–400. https://doi.org/10.1353/mod.2000.0056.

Fink, E. (1968 [1957]): "The Oasis of Happiness: Toward an Ontology of Play".*Yale French Studies*, no. 41, 19. https://doi.org/10.2307/2929663.

Fink, E. (1979): Grundphänomene Des Menschlichen Daseins. Freiburg: Alber.

Fuchs, M. (2017): "Ruinensehnsucht: Longing for Decay in Computer Games". Transactions of the Digital Games Research Association 3 (2), pp. 37- 56. https://doi.org/10.26503/todigra.v3i2.68.

Goethe, J. W. (1970): Italian Journey: = [1786-1788]. London: Penguin Books.

Gualeni, S. (2015): Virtual Worlds as Philosophical Tools: How to Philosophize with a Digital Hammer. New York: Palgrave Macmillan.

Gualeni, S. (2016): "Self-Reflexive Videogames: Observations and Corollaries on Virtual Worlds as Philosophical Artifacts". GAME. Game as Art, Media, Entertainment 1 (5).

Gualeni, S./Vella, D. (2020): Virtual Existentialism: Meaning and Subjectivity in Virtual Worlds. Cham: Palgrave Macmillan.

Gualeni, S/Fassone, R. (2023): Fictional Games: A Philosophy of Worldbuilding and Imaginary Play. Bloomsbury Academic.

Heidegger, M. (1983 [1953]): Einführung in Die Metaphysik. In Jaeger, P. (eds.) Gesamtausgabe 40. Frankfurt am Main: Klostermann.

Heidegger, M. (1993a [1951]): Building, Dwelling and Thinking. In Krell, D. (eds.) Basic Writings: From Being and Time (1927) to The Task of Thinking (1964). San Francisco, Calif.: HarperSanFrancisco, pp. 343-364.

Heidegger, M. (1993b [1954]): The Question Concerning Technology. In Krell, D. (eds.) Basic Writings: From Being and Time (1927) to The Task of Thinking (1964). San Francisco, Calif.: HarperSanFrancisco, pp. 307-342.

Heidegger, M. (1995 [1930]): The Fundamental Concepts of Metaphysics: World, Finitude, Solitude. Bloomington: Indiana university press.

Heidegger, M. (1996a [1953]): Being and Time. Albany, NY: State Univ. of New York Press.

Heidegger, M. (1996b [1942]): Hölderlin's Hymn The Ister. Bloomington: Indiana university press.

Heidegger, M. (2009 [1921]): Phenomenological Interpretations of Aristotle: Initiation into Phenomenological Research. Bloomington: Indiana University press.

Henderson, J. V. (2010): "Cities and Development". Journal of Regional Science 50 (1), pp. 515–40. https://doi.org/10.1111/j.1467-9787.2009.00636.x.

Hollis, L. (2013). Cities Are Good for You: The Genius of the Metropolis. London: Bloomsbury Publishing.

Husserl, E. (2011 [1893 – 1917]): On the Phenomenology of the Consciousness of Internal Time *(1893-1917)*. Dordrecht: Kluwer Academic Publ.

Husserl, E. (2012 [1913]): Ideas. New York: Routledge. https://doi.org/10.4324/9780203120330.

Husserl, E. (2015 [1901]): Logical Investigations vol. I. New York: Routledge.

Johnson, M. (2014): Lives in Ruins: archaeologists and the seductive lure of human rubble. New York: HarperCollins Publisher

Kurir Mateja. (2022): "On Home (Das Heim) and the Uncanny (Das Unheimliche)". Heidegger'. Phainomena 31 (120–121), pp. 121–45. https://doi.org/10.32022/PHI31.2022.120-121.6.

Lee, V. (1925): The Golden Keys and Other Essays on the Genius Loci. Leipzig: Bernhard Tauchnitz.

Mackay, L./Roberts, C. (2023): "Peace in the Valley: A Media and Discourse Analysis of Eric Barone's *Stardew Valley* Through Utopian Theory". Games and Culture 20 (1), pp. 3–19. https://doi.org/10.1177/15554120231187793.

Manca, M. (2020): "Rovine. Una Tipologia Ed Un Tentativo Di Definizione". Filosofia e Robotica, Filosofia, LXV, pp. 145–60.

Murchadha, F. (2002): "Being as Ruination: Heidegger, Simmel, and the Phenomenology of Ruins". Philosophy Today 46: pp. 10–18. https://doi.org/10.5840/philtoday200246Supplement2.

Norberg-Schulz, C. (1980): Genius Loci: Towards a Phenomenology of Architecture. New York: Rizzoli.

Plunkett, D. (2021): "Technology, Dwelling, and Nature as "Resource": A Reading of (and Some Reflections on) Themes from the Later Heidegger". Inquiry, pp. 1–71. https://doi.org/10.1080/0020174X.2021.1989030

Resta, C. (2020): "La Legge Segreta Della Migrazione". In Casu, M. (eds), *Was Heißt Stiften? Heidegger Interprete Di Hölderlin*, Roma: Istituto Italiano di Studi Germanici, pp. 115–131.

Ricoeur, P. (1988 [1983]): Time and Narrative, Vol 3. Chicago: University of Chicago Press.

Ruin, H. (2012): "Thinking in Ruins: Life, Death, and Destruction in Heidegger's Early Writings". Comparative and Continental Philosophy 4 (1): pp. 15–33. https://doi.org/10.1558/ccp.v4i1.15.

Ryan, M. (1991): Possible Worlds, Artificial Intelligence, and Narrative Theory. Bloomington: Indiana University Press.

Sennett, R. (2018): Building and Dwelling: Ethics for the City. New York: Farrar, Straus and Giroux.

Simmel, G. (1958 [1911]): The Ruin. In "Two Essays". The Hudson Review 11 (3), pp. 379–85. https://doi.org/10.2307/3848614.

Soja, E. W. (2010): "The City and Spatial Justice". In Bret, B./ Gervais-Lambony, P. (eds.) Justice et injustices spatiales, pp. 56–72. Nanterre: Presses universitaires de Paris. https://doi.org/10.4000/books.pupo.415.

'Symbol'. n.d. In *Etymology*. Accessed 30 November 2024. https://www.etymonline.com/word/symbol.

Tolia-Kelly/Waterton, E./Watson, S. (eds.) 2018. Heritage, Affect and Emotion: Politics, Practices and Infrastructures. Critical Studies in Heritage, Emotion and Affect. London: Routledge, Taylor & Francis Group.

Tufo, G. (2008): "Sulle Orme Dei Banditi e Dei Briganti Del Territorio Arnate". In Perugia. Tesori Nella Campagna. Perugia.

Tufo, G./Bastianelli, G. (1998): Notizie e Ricordi Spettanti Alla Chiesa Di S. Emiliano Di Ripa. Perugia.

Vella, D. (2010): Virtually in Ruins. The Imagery and Spaces of Ruin in Digital Games. Dissertation, Malta: University of Malta.

Vella, D. (2015): 'The Ludic Subject and the Ludic Self: Analyzing the 'I-in-the-Gameworld'. Copenhagen: ITU Copenhagen.

Vella, D. (2016): "The Ludic Muse: The Form of Games as Art". Counter Texts 2 (1): 66–84.

Vella, D. (2019): "There's No Place Like Home: Dwelling and Being at Home in Digital Games", In Aarseth, E./ Günzel S. (eds.) Ludotopia: Spaces, Places and Territories in Computer Games, pp. 141–66.

Zahavi, D. (2005). Subjectivity and Selfhood: Investigating the First-Person Perspective. Cambridge, Mass: MIT Press.

Table of figures

Figure 1. Aerial picture of Ripa. 1970. Courtesy of Associazione Federici, ProArna, Biblioteca Federici.

Figure 2. Ripa community centre. Left. 1934. Probably Afterwork Association during fascism. Right: 1958 Headquarters 12 Borough of Arna. Courtesy of Associazione Federici, ProArna, Biblioteca Federici.

Figure 3. The map of Pelican Town. Screenshot by the author.

Figure 4. Community centre. Screenshot by the author.

Figure 5. Community centre, inside. Screenshot by Author.

Ruins as Image

Moving Image. Changing Landscape

Trevor Borg

Abstract

The current article considers how the moving image intersects with the changing landscape and investigates how that crossing might raise questions about our relationship with place and space. Central to the discussion is Olea, a visual, auditory, and olfactory installation exhibited at the Malta Biennale 2024. The work features a 'mythical' figure situated within an indeterminate temporal realm that spans across historical epochs. The speculative narrative brings up a multiplicity of contexts whereby Olea may (re-)connect with the past by engaging with ancient ruins and artefacts in various conjectural settings, imbricating her personal and familial memories with collective ones. Drawing on phenomenological experience, the installation immerses the viewer in an intimate, multi-sensory space, enriched with layered visual and sonic elements. The condensed environment is intended to elicit a sense of urgency, provoking action in response to the rapidly changing landscape. The piece navigates between temporal dimensions, bridging past and present, through digital media. An implicit poetic-political interweaving serves as a central axis, fostering diverse interpretative pathways and inviting viewers to reimagine alternate scenarios through their own unique perspectives.

Keywords

Moving Image, Ruins, Culture, Landscape, Installation

The Case of Olea

Contemporary society inhabits an ambivalent space, one that is imbued with cutting-edge technology and forward-looking aspirations, blurring the physical with the virtual, and at the same time falling to pieces due to environmental degradation, climate change and conflict. As humanity sets its sights on colonising distant planets, it risks accelerating the collapse of Earth. The world we are abandoning may, in a few decades, resemble the ones we aspire to inhabit – an outcome of our own actions. It could be argued that our planet is in a ruinous state of degradation; many regions now bear little resemblance to what they were only a few centuries ago. The landscape has been disrupted to the point of near irreversibility. Yet, rather than prioritising the conservation of what we have carelessly neglected, our focus has shifted outward, as we seek solutions far beyond our own planet. This could be the spark of new hope – or the quiet trace of hope slipping away.

This article takes Olea's position as a point of departure to reflect on the intersection of past and present and to speculate on possible futures. The installation unfolds symbiotically through political and poetic narratives, positioning itself as one among many critical voices. The past is emphatically examined to provide deeper insight into the present, allowing for a nuanced interpretation of contemporary realities through historical reflection. We uncover the ancient past through ruins and fragments – unearthed objects layered like palimpsests within the strata of the earth. These remnants allow us to reconstruct, both conceptually and historically, what has been dismantled, erased, or disrupted. Ruins, however, may reveal more about the present and future than the past, offering us a glimpse into our current trajectory. Through remnants we gain insight not only into what has been but also into where we may be heading. Ruins stand as haunting monuments to what once was, embodying the fragility and impermanence of human endeavour. They are spaces where time seems suspended, allowing us to glimpse the past while confronting our present vulnerabilities, and shortcomings. These landscape features may be intentional or incidental and are not always the result of indifference to what once was. Yet, how we engage with ruins reveals much about what may come.

For Malta Biennale 2024, I created an installation bringing opposing temporal aspects together to create a speculative, multi-layered narrative that belonged to the present and the past and to neither. The narrative began to unfold in an in-between space, where the past and present fluctuate and exchange positions. 'Olea' serves as both the title of this work and the name of the central figure who inhabits it. The word *olea*, derived from Latin and commonly associated with the olive tree, forms part of the scientific name *Olea europaea*. Olive oil production in the Mediterranean boasts a rich, extensive history spanning thousands of years and remains an important cultural and economic activity in the region. On the small Mediterranean island of Malta, various archaeological sites consisting of an architectural complex associated with ancient olive oil production have been discovered, while many other finds related to oil production were uncovered across the Maltese islands (Anastasi/Vella 2018: 276-277). Many of the discoveries are believed to be of Roman origin, suggesting that oil production was prominent during that period. While the Romans did not initiate the olive oil industry, they greatly expanded and industrialised it. The Roman olive oil industry was a cornerstone of the ancient economy and the production and distribution of olive oil was vast and sophisticated, reflecting the importance of this commodity in various aspects of Roman society.

The phantom-like Olea walks with the grace of a timeless being, a figure draped in the ambiguities of countless eras. Her footsteps are soft, yet each one reverberates through the fabric of time and space, leaving marks on the ground.

The installation integrates moving image, sound design, and the aroma of olives, creating a complex interplay of sensory stimuli within the cinematic narrative. A multi-layered composition that, to borrow London's (2020: 159) words, consists of "complex content and exacting uses of space".

Both diegetic and non-diegetic sounds are employed to underscore key elements while complicating others, imbuing the narrative with nuanced layers that carry a delicate, almost fragile tension. This tension serves to maintain the plot's precarious equilibrium, subtly amplifying its dynamism and depth, echoing what Chion (1994: 5) defines as 'added value'; a term that describes the critical relationship between visual and auditory elements in shaping the viewer's perception. The immersive 5.1 soundscape resonating within a confined space encapsulates multiple historical moments, allowing the viewers to explore different periods, events, and narratives embedded in a single location.

The images are haunted by ruins emerging from the past right into the present. They act as sites of memory, an appreciation of continuity and change in a specific place. A subtle olive hue permeates the sequence of images, contrasting with the manipulated industrial sounds that fill the exhibition space. Olea's gaze – a deep green reflecting the olive leaves recurring throughout the journey – carries traces of millennia. In the heart of the present, Olea finds an overpowering sense of connection, a recognition of the threads that weave the tapestry of time. Her passage through time and space becomes a journey of reflection and renewal, enabling a profound reconnection with the past.

The physicality of ruins creates an aesthetic experience that influences how people perceive and emotionally connect to a place. This material decay can provoke contemplation on concepts like mortality, resilience, and the passage of time, which are central themes in phenomenological analyses of place. Some shots feel like excerpts or fragments, offering glimpses rather than the entire narrative – selected moments rather than a complete film. These moments serve as ruins in their own right – remnants of a larger narrative, condensed and decontextualised, much like artefacts displayed in museums. Ginsberg (2004: 238) argues that this is akin to something extracted from its original context and placed before us for contemplative enjoyment; an excerpted piece feels constrained, as if time itself is compressed. This instils a layer of urgency within the work, leaving viewers feeling either helpless or compelled to act, however, the precise nature of the action remains deliberately ambiguous. The work is intended to remain open and to freely oscillate between the poetic and the political.

Ginsberg (2004: 221-222) tells us that although ruins are often associated with architecture, they may also exist in other forms, including the visual arts. Ginsberg discusses how fragments of objects in museums allow us to speculate and imagine what and how they might have been. Olea reterritorialises archaeological ruins and objects retrieved from ruinous sites into a video installation; it works in time-based media to juxtapose the ancient with the present, allowing us to go back in time to conceptually encounter that which is gradually fading away

(Figure 1). The ghostly figure connects us with our cultural past; it invites us to re-collect lost fragments and to piece together narratives of our own. This connection is forged by means of ancient artefacts she discovers lying on the shore – possibly the remnants of a Roman-era shipwreck. The work thrives on uncertainty and speculation, revealing nothing at surface level. Instead, viewers are invited to decipher the layers concealed beneath the visibly exposed, encouraging a deeper engagement with what lies beneath.

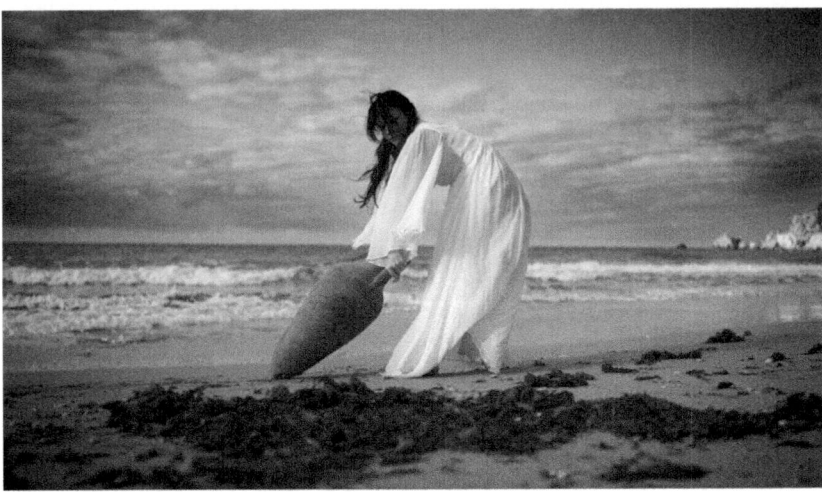

Fig. 1: Still-shot from Olea, *(2024)*

My place-oriented practice often gravitates toward the phenomenological, focusing on experiences that engage the body and senses directly. Olea's story unfolds through her movement across the landscape; she herself performs the act of walking. Her walk not only propels the images forward but also introduces more material into the narrative, as walking itself is a way of knowing (Ingold/ Vergunst 2016: 5). Places are encountered in a non-linear unfolding as the central figure, without ever uttering a word, moves us through space and time. There is no beginning or end to this walk – it keeps on going, it makes time reversible (Gros 2009: 128). Ruins are not merely remnants of physical structures but symbols that evoke the passage of time, human experiences, and social transformations. As Olea moves forward, she carries with her the weight of history – a timeless figure within an ever-evolving world, her path serving as a symbolic bridge between what has been and what is yet to come. Her connection to the trees is profound; their gnarled trunks and sun-kissed branches stand as silent witnesses to her timeless passage. The installation's lyrical layers counterpoint its underlying political and ecocritical tones, which emerge more fully through a closer, more nuanced reading of the work.

Placing Ruins

Ruins are positioned within the cultural, historical, and spatial context of a landscape and they should be considered not merely as isolated fragments but as integral parts of a place's ongoing narrative. *Olea* was filmed in four different locations in Malta, two of which are directly linked to the Roman period. References to antiquity in the work are subtle, requiring a discerning eye and familiarity with the island to uncover their exact location. One can discern a 'virtual' element in this time-based work which "requires the mutual enfolding of here and elsewhere, and now and then" (King 2015: 70); a looping of time and place that allows the viewers to approach or abandon the work at any point or to watch the images unfold repeatedly. This might complicate the engagement of the audiences with the work and the piece could come across as disorienting, however, the so to speak 'sequential' disruption is done intentionally to allow the myriad narratives to jostle for space. The audience is encouraged to virtually 'edit' the sequence, starting and ending the viewing at any point in time to encourage alternate and unlikely readings.

Featured locations comprise Il-Mixquqa (Golden Bay), a sandy beach in the Northern part of Malta not far from a complex of Roman baths, San Pawl Milqi in Burmarrad, where extensive Roman oil production architecture (and other non-Roman remains) has been uncovered, and the Domus Romana, a museum built on the site of a wealthy Roman residence featuring intricate mosaics still in situ. The entwining of nature and culture, especially in sites exposed to the natural elements, reveal the un-cultivated beauty of decay amidst new growth. Ruins reflect natural and cultural processes of decay and renewal, embodying the cyclical nature of human existence. In place-oriented practice, this symbolism offers a framework for understanding ecological and social transformations, reminding us of the impermanence of human constructs and the resilience of the natural environment. The ruinous sites are framed in between expansive views of the open sea and a landscape of olive trees within the designated national park at Ta' Qali. Deep underwater scenes and aerial images of the landscape feature sporadically in between cuts. These are mostly establishing shots intended to gradually reveal places from afar and to combine them together to further blur any delineations that might explicate a change in place. This encourages the audience to lose itself and, in the process, allowing the 'unfamiliar' to appear, discovering new places as multiple narratives unfurl and become interwoven (Solnit 2017: 22).

Harbison (2015: 184) argues that damage, whether as form or anti-form, holds the capacity to reveal certain truths. He illustrates this with film surfaces that are deliberately attacked, scratched, and worn, creating a novel sensory experience through the aesthetic of destruction. In the case of *Olea*, the film is disrupted, dislocated and fragmented, merging disparate locations into a unified whole to reconstruct a narrative based on ruins dispersed across various parts of Malta. The objective is to create an experience distinct from that of viewing ruins in their so-called 'original' settings, as one might encounter in documentary films. Here, historical accuracy is not important; instead, the focus shifts toward presenting alternative realities that engage more deeply with contemporary concerns than with past narratives. There is nothing obvious or direct in this piece as every component is loaded with cues that necessitate close (re-)readings.

The 'restrictive' space hosting the installation bordered on claustrophobic, with Olea's figure emerging like a classical deity – larger than life and almost intimidating to the audiences. A tailor-made room constructed from panels and soundproof material, measuring eight meters wide, five meters long, and three meters high, was erected at the National Art Museum in Valletta, to accommodate the installation. The matt black walls did not allow for any light spills or reflections except for one of the longer walls onto which the video was projected (Figure 2). Black flooring was installed to camouflage the surrounding walls and to obfuscate the edges and the perimeter of the room. The dark interior aimed to instil an uneasy feeling and a sense of urgency, creating an atmosphere that does not encourage one to stay for longer than needed. As embodied beings, we inevitably form relationships with the places and spaces we inhabit (Trigg 2012: 1). Our perception of the surrounding space shapes our relationship with the places encountered in the room that functions like a heterotopia – a transient single place that juxtaposes multiple spaces and sites (Foucault 1984). Viewing the images inside an enclosure is not like viewing images on television, in a cinema or on a mobile phone or tablet. The immersive space is intentionally designed to orient our bodies in the places that are encountered within it, as if we are walking through the landscape and among the ruins with the figure herself. The anxiety-inducing space contrasts sharply with Olea's movement, as she drifts like a weightless feather, seemingly untouched by the passage of time. This dissonance prevents the viewer from settling in one place, instead prompting them to keep moving, to engage actively, and to search for clues and meanings.

Doreen Massey (2005: 71) claims that places are not coherent but act as foci for the meeting and nonmeeting of occurrences that were previously unrelated; this, she argues, generates novelty. The piece encourages correlations between distinctive places, including places within which we can already discern gradations spanning millennia, as in San Pawl Milqi. In one of the scenes, the figure can be seen entering the ruins of a Roman period agricultural villa through a 17th century chapel located on the same site. Similar time-place juxtapositions occur throughout the piece, challenging the linear chronology of historical narratives. Trigg (2012: 7) argues that time manifests itself in place mainly through 'movement' and 'stasis', two significant factors that are distinctly felt when experiencing this artistic installation. As the figure traverses the room, entering from one side and exiting through another, the viewers are compelled to remain motionless, ensnared within the enclosed space imbued with the scent of olive oil. Scent plays a profound role in defining place by engaging the sensory experience, evoking visceral, immediate responses, drawing on deep-seated memories and associations. It acts as a marker that grounds individuals in their surroundings, creating a sense of belonging or dissonance based on familiar or unfamiliar smells. The scent of olive oil can evoke cultural, historical, or even personal connections to Mediterranean landscapes, rituals, and traditions. It can influence how a space is navigated, remembered, or even experienced, and it can reinforce a sense of temporality, marking an environment that changes according to the passage of time.

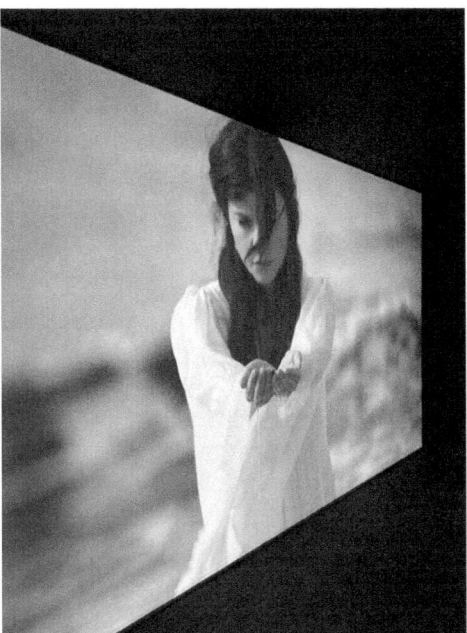

Fig. 2: *Inside the enclosure*

The installation creates an uncanny space, where nothing is fixed or immediately apparent. Places are continuously formed and reformed throughout the journey, intertwining seamlessly until the boundaries between them become indistinguishable. This instability amplifies the ruinous atmosphere conveyed through the moving images, deconstructing both place and time in an unbroken loop.

"This movement of perpetual beginnings is one of phenomenology's great strengths. By leaving the world exposed to uncertainty, dynamism is maintained and our own place in that tension is amplified." (Trigg 2024: 41)

Yi-Fu Tuan (1977: 6) suggests that while space is associated with movement, place requires a pause. In *Olea*, pauses punctuate the walk, inviting viewers to reconstruct their own place(s). As Tuan notes, it is through these pauses in movement that place is meaningfully created. Such pauses in movement across the landscape encourage the audiences to imbue the created places with personal and collective meanings. Place emerges from human experiences, memories, and cultural practices associated with a landscape; therefore, we need time to make place. The video and audio installation is looped to enable smooth, seamless re-viewing, aided by the softening of edges within the room that contribute to making the space more intense. Each re-viewing extends the time for placemaking, encouraging 'dwelling' and fostering a heightened sense of being, as suggested by Ingold (2011: 114).

The images unfolding within the custom-built room juxtapose smooth with striated space, whereby a non-homogenous unlimited space is continuously reconfiguring the edges of the enclosure (Deleuze/Guattari 2004). A smooth space is homogenous, characterised by shifts in directions and infinite linkages, while striated space is defined as a conventional homogenous whole (Lorraine 2005: 253-4). This uncanny configuration introduces a fragile tension that unsettles the apparent balance evoked by Olea's gentle demeanour as she invites the viewers to walk with her across time and space.

"No sooner do we note a simple opposition between the two kinds of space than we must indicate a much more complex difference by virtue of which the successive terms of the opposition fail to coincide entirely. And no sooner have we done that than we must remind ourselves that the two spaces in fact exist only in mixture: smooth space is constantly being translated, transversed into a striated space; striated space is constantly being reversed, returned to a smooth space." (Deleuze/Guattari 2004: 524)

Deleuze and Guattari (2004: 539-40) argue that space can escape the limits of striation; it only entails a small deviation, or as they call it a declination, to escape it. The installation situates ruins within 'other' places, blurring the boundaries between the past and the present, between here and there. Fixed delineations give way to fluid declinations, enabling the emergence of multiplicities and alternative contexts, where meanings unfold through rhizomatic interconnections. The moving images function as an open window, inviting the viewer to appropriate the places 'outside' and transform them into new spaces within. Deviations encourage new stories and as Haraway (2016: 35) argues, "[i]t matters what stories tell stories." Ruins are not static; they are constantly subject to reinterpretation; they often carry symbolic significance acting as focal points for collective memory and as spaces where local identity is negotiated and reaffirmed. The artistic work seeks to problematise historicity through "real stories that are also speculative fabulations and speculative realisms" (2016: 10). Ingold (2011: 208) argues that "[s]tories help to open up the world, not to cloak it." Fabricated realities enable the emergence of smooth space, where fact and fiction intertwine, allowing ruins to transform into future places. We should refrain from viewing ruins merely as memories of the past and instead see them as maps guiding us into the future, perhaps helping us avoid many of the anthropocentric paths we have mistakenly followed in recent and not so recent times.

Ruins as (Poetic/Political) Aesthetic Experience

Ruins are often appreciated for their romantic allure and nostalgic connection to a lost past; however, they also embody potent memories and reveal capitalism's relentless drive to reclaim sites in favor of a profit-oriented future. Artists have always been fascinated by ruins. Giorgio de Chirico incorporated ruins into his work to evoke a sense of timelessness and mystery, using them as symbols of both decay and permanence. These architectural remnants often appeared in his hypothetical landscapes, where they played a crucial role in creating a surreal atmosphere that blurred the boundaries between past and present. Land artist Robert Smithson investigated decay and entropy in his work, revealing the transformative effects of time on structures and materials. Photographic duo Bernd and Hilla Becher, for fifty years, documented industrial architecture and buildings on the verge of obsolescence before decaying and vanishing from the landscape. Their work, focused primarily on the aesthetics of form and function, serves as a visual record of structures likely bound to fade and decay. German artist Anselm Kiefer creates sculptures and paintings that embody decay within their very materials, merging the poetic and the political as underlying elements of ruination. The aesthetic experience of ruins is a complex topic deserving its own comprehensive study, encompassing not only the visual arts but also film, literature, video games, theatre, music, and performance.

DeSilvey and Edensor (2012: 467) maintain that interpretations of ruins often remain narrowly focused on visual aesthetics and certain philosophical constructs, frequently overlooking critical aspects of power dynamics and the intrinsic presence of these decayed spaces. The concept of 'creative ruins' is especially relevant against a global backdrop marked by conflict and the widespread ruination of economies and ecologies; they argue that "[w]hile there may be an emergent political sensitivity in the attention paid to ruins, this is always undercut by the potential for other interpretations and appropriations." (ibid: 471) Sometimes, decay possesses a romantic and aesthetic beauty that cannot be overlooked, and an artistic approach may be one of the various ways to underscore its political dimensions.

Ancient and contemporary ruins constitute a poetic and political intertwining that goes beyond the structural. Ruins embody a dynamic critical power that is contingent upon their historical and cultural context that changes over time (DeSilvey/Edensor, 2012: 469). In sites shaped by ruination we encounter an archaeology of remnants – what endures and what has vanished. Archaeologists operate at the threshold between presence and absence, meticulously peeling back layers of earth and reading the landscape (Vergunst 2016: 35). Through the installation, visitors become archaeologists themselves – exploring, unearthing, and piecing together fragments to construct their own personal narratives. Rancière (2004: 35) writes that "[p]olitics and art, like forms of knowledge, construct 'fictions', that is to say *material* rearrangements of signs and images, relationships between what is seen and what is said, between what is done and what can be done." [original emphasis] This emphasises the fact that art and politics actively shape perception and action; creating new ways of understanding, experiencing, and transforming things.

"The political is not just about people, rights and relationships; it is about things too" according to Pearson and Shanks (2001: 50). A particular artefact consisting of an amphora appears in various contexts throughout the installation; a material object serving as the interface between archaeology and culture (2001: 33). Olea is depicted dragging an amphora – a vessel historically used in antiquity for the storage and transport of commodities, especially liquids such as olive oil – from the seashore into an abandoned structure. Her actions suggest an attempt to rescue and secure the artefact, safeguarding it from potential theft or destruction. In an interview with Daniel Xerri for *MaltaToday*, I discussed potential interpretations of these scenes, offering insights into their possible meanings and suggestions:

"We all know that Malta is like an open museum. There are ruins and remains everywhere; wherever you dig and search, you're going to find something. This is a blessing but also problematic. Given the size of the island, space is very limited and so a lot of tension is created when it comes to new development.
We are doing an excellent job in protecting the artefacts and cultural elements that we have inside our museums. We're constantly investing in new museums and new technologies to make collections more accessible to the public. However, there are many open sites that are facing different threats and we are at imminent risk of losing them once and for all." (Borg interviewed by Xerri, 2024)

I have frequently incorporated both authentic and fabricated artefacts into my work. In 2019, I exhibited *Cave of Darkness: Port of No Return* in the Arsenale at the Venice Biennale. This installation, comprising hundreds of sculptures made from pottery, stone, wood, animal bone, and 3D-printed replica bones of extinct mammal remains from a cave discovery in Malta, created an imagined museum that blurred the boundaries between history and fiction. Morgan (2019: 97) described this work as provocative since it questions the urgency of a natural and cultural collapse for which humanity is responsible. More recently, I have shown *FIND:FADE* (2024), consisting of a contemporary cabinet, reminiscent of the *Wunderkammer*, housing various real and replica artefacts. The work is intended to inquire about historicity and how it is affected by time, as stories are retold, forgotten, erased and revised. Rancière (2004: 35) maintains that "[w]riting history and writing stories come under the same regime of truth." He challenges the traditional distinction between historical writing and storytelling, arguing that both are constructed through narrative approaches and are shaped by the same underlying principles of representation and interpretation.

Olea is not anchored to any particular era, embodying an atemporal quality. The figure appears amidst ruins, yet there is no intent to confine it to a specific historical period. Ancient and later ruins are juxtaposed in a non-chronological arrangement, creating a timeless narrative. The figure itself could belong to any epoch, existing outside the boundaries of a defined historical timeline. The lingering olive scent within the chamber, emitted at specific intervals, enriches the engagement with the work. An aroma revealing a complexity beyond the immediate, with notes that capture the subtleties of ripening fruit, the mineral richness of the soil, and hints of salt carried from the nearby sea. Overlaying the visual and olfactory experience is a meticulously produced 5.1 surround audio composition, transforming captured sounds into an industrial soundscape. This auditory layer, manipulated and remixed to abstract familiarity, envelops the audience, creating an immersive environment that oscillates between the organic and the mechanical. The result is a soundscape that underscores the intersection of nature and human intervention, evoking a sense of both the land's primordial rhythms and the modern machinery that now shapes it. This juxtaposition challenges the audiences to engage with the space not only as a sensory experience but as a conceptual exploration of the evolving landscape. Rancière (2011: 56) argues that the new sensory fabric originating in an artistic production draws upon ordinary experience to make the piece more accessible to the audience. *Olea* invites audiences to engage with ruins in an embodied way. These consist of what Lucas defines as slow ruins, sites that slip into ruination gradually as they become incrementally abandoned (DeSilvey/Edensor 2013: 466-67). The dark designated space encourages the viewer to enter an unexplored territory. It encourages discovery. There is a lingering dissonance hovering within the space generated by the industrial soundscape that permeates each and every image.

Editing and colour grading play a pivotal role in shaping the visual aesthetics and influencing audiences' perception. There is no script and neither a voice over here, as the storytelling relies on Olea's trajectory and the multiple intricate layers of processed audio. The overall aesthetic is generally unsaturated but a tinge of green can be discerned throughout parts of the sequence. I often approach colouring in the same way I approach my paintings, mixing colours in post-production as I would mix paint in my studio. I find that this approach imbues the image with an organic fabric, more gradated and sensitive to the shades and tones in between, where some of the storytelling elements develop and open out.

The piece remains vague throughout; it thrives on uncertainties and conjecture and like the landscape itself it is difficult to contain. As Olea urgently pushes an amphora across the shore and through multiple other places, more questions arise regarding the significance of this ambiguous gesture. We are confronted with images of the deep blue sea, amorphous and indefinite, and of a mythical figure stranded on an island; uprooted and disoriented, she wants to reconnect with her past in an attempt to shape the present. She desperately engages with objects that might help mitigate the feeling of dislocation and situate herself and the remnants of her ancestors in a more secure place. But this is all conjecture, as nobody really knows the story; it is bleak and hazy but that is not going to discourage us from writing our own. Landscape can be deciphered as text; it presents itself as a space within which we may find or lose ourselves (Mitchell 2002: 1-2). Olea's story is also ours and this makes us sympathetic to her cause. She directs our gaze, we follow her with our eyes, and by trying to understand what is going on, we are gradually drawn back and forth into her story as we introduce our own text and relate it to hers. The meaning-making process is a process where the ruptures in the narrative are completed and extended, allowing us to act not merely as spectators but as active participants and narrators.

Mitchell (2002: 20) argues that "[l]andscape is now more precious than ever– an endangered species that has to be protected from and by civilization". What are the implications of a rapidly changing landscape? The work seeks to highlight these doubts and cultivate additional uncertainties on multiple levels, thereby prompting further inquiry and reflection.

"The unstructured exploration of possible pasts, and the encounter with involuntary memories, can perhaps occur more readily in ruins that remain 'open' – managed lightly, if at all, still caught up in dynamic processes of decay and unmaking. However, as we have suggested already, this liminal state is actually a fragile and ephemeral achievement. In some places, where economic depression militates against inward investment, ruins may linger for decades; in sites of more dynamic social and economic change, ruined structures are apt to be swiftly razed, reclaimed or restored." (DeSilvey/Edensor 2012: 472).

Tsing (2015: 211) argues that "[t]he effect of industrial ruins on living things depends on which living things we follow." We often view the remnants scattered across landscapes through an anthropocentric lens, overlooking the fact that humans are not the only inhabitants of this world. As Tsing (2015: 211) points out, ruins – particularly industrial ones – can serve as refuges for certain species, such as insects and parasites, while being destructive to others. Beyond the ruins of human structures, we must turn our attention to the ecological sphere, where the extinction of species is wreaking profound damage. If this trajectory is not reversed, the world itself risks descending into a state of ruin. "The extinction of a critical number of species would mean the destruction of long-evolving coordination and interdependencies." (Tsing/Swanson/Gan/Bubandt 2017: 4) Our environmental degradation threatens to unravel the delicate interconnections that uphold the balance of nature, leaving behind a landscape not of history's remnants, but of irreversible loss.

The landscape surrounding ruins can be as revealing as the ruins themselves, as it preserves both the original context and the transformations that have occurred over time. When ruins are decontextualised, original narratives are lost, and new ones emerge – stories that speak to the future of the landscape and the powerful forces shaping it. The installation situates the audiences in an in-between place. As Casey (2002: 35) argues, to be in-between is not to be placeless but to be in a place in a formative way, in a place between other places, an interplace. This is a privileged vantage point from which one can observe both the vanishing and the emerging. Olea's relentless drive to salvage the amphora can be seen as a symbol of the urgent need to preserve what is gradually being lost to the advancing process of disintegration.

The Shape of Ruins

Macfarlane (2016: 12) writes that the word landmark is derived from the old English word *landmearc* which means "an object in the landscape which, by its conspicuousness, serves as a guide in the direction of one's course". Similarly, ruins offer insight into the direction – or lack thereof – in which a country is perceived to be progressing. This applies not only in archaeological or historical terms but also on ecological, cultural, and even economic levels. Merleau-Ponty (2004: 70) emphasises that it is impossible to separate things from the way they appear – remnants speak of place and tell us about the people who make and unmake it. By examining remnants in a given place, we learn about the people who interacted with them; their values, their practices, and their transformations of the environment. The material remains of human presence in a place thus become an embodied memory, offering insights into how a place was lived in, altered, and understood over time. In this way, remnants carry the potential for reimagining and reinterpreting the ongoing relationship between people and the places they inhabit.

Walking through ruins, both physically and conceptually, is a challenging endeavour, as Edensor (2016: 127) notes; such walks defy a regular rhythmic gait due to their inherent unpredictability. The custom-built room housing *Olea* offers a similar experience, where the audience is compelled to reflect on each step, as the space itself remains unpredictable and unknown. Edensor (2008: 129) describes this as "unfamiliar movements in unfamiliar spaces" that evoke a heightened sense of awareness. The concept of ruin-awareness opens up a space for further exploration and consideration of the delicate interplay between decay and preservation, absence and presence.

Ruination takes on many forms, often hidden or camouflaged to the extent that we may not even recognise it as such. Highly industrialised and commercialised spaces may, in fact, rest atop layers of past ruins – material, economic, and cultural. In European cities, we catch fleeting glimpses of ancient ruins while riding the Metro, or through glass floors in shopping malls that reveal buried remnants of earlier civilisations. Yet, the mall itself may also be in economic ruin, an instability invisible until a backstory emerges in the headlines. Ruins are often deceptive, concealing histories of decline beneath façades of progress and modernity, challenging us to discern what is thriving from what is quietly decaying. Olea moves cautiously through space and time, fully aware that each step could disturb the landscape and unveil unpredictable revelations. Ruination is located across space and time, and these are far from being immutable as they are very much subject to historical change (Huyssen 2003: 24). Thus, the present, though immediate and tangible, cannot claim to be an unmediated truth of what was, nor can it be the sole compass by which we navigate what may be.

This article aims to reconstruct an architecture of ruins through the integration of moving images, audio, and other objects. The installation's layered spatial arrangement allows for a dynamic interplay of meanings, where different interpretations can coexist and interact, generating a complex assemblage within the same immersive environment. The objective is to stimulate further discourse on the prominent role of ruins within the cultural, historical, and political landscape, thereby encouraging alternative approaches to engaging with them. *Olea* constitutes a montage; a composite work created by assembling and repurposing different images into a single composition. Pearson and Shanks (2001: 52) argue that the montage appropriates and borrows from different contexts, similar and dissimilar, with the aim to construct something novel, to achieve new meaning and understanding. Instead of depicting subject matter as it is, this bringing together disrupts and interrupts representation, encouraging new forms of discovery (ibid: 52).

Olea highlights the dynamic interplay between the moving image and the landscape, using a phenomenological approach to create an immersive, multisensory experience. By situating a 'mythical' figure within shifting temporal contexts, the installation invites viewers to reflect on their own connections to place. Through its poetic and political layering, *the work* fosters multiple engagements, encouraging different meanings in response to our evolving landscapes. It creates a multilayered space in which ruins can be encountered and explored, enabling their conceptual reconstruction. This process is facilitated by the audiences, who bring their own understandings and interpretations to the narrative, enriching the work with diverse perspectives and significance. It operates on multiple interpretative levels, granting audiences the freedom to engage with it through their preferred lens. There is no fixed or singular way of viewing the piece; rather, it continuously unfolds and reconfigures itself with each subsequent encounter, inviting new readings and meanings over time. *Olea* opens up ambiguous avenues for engaging with ruins, not through a historical perspective, but on a closer and more personal level. By venturing into uncharted realms, the work invites exploration and the discovery of new trajectories, permeating the encounter with different prospects.

References

Casey, E. (2002): *Representing Place*. Minneapolis: University of Minnesota Press.
Deleuze, G./Guattari, F. (2004): *A Thousand Plateaus*. London: Continuum International Publishing Group.
DeSilvey, C./Edensor, T. (2012): "Reckoning with Ruins." *Progress in Human Geography* 37(4), pp. 465-485. https://doi.org/10.1177/0309132512462271
Edensor, T. (2016): "Walking Through Ruins." In: T. Ingold/J. L. Vergunst (eds.), *Ways of Walking*. London: Routledge, pp.123-141.
Foucault, M. (1984): "Of Other Spaces: Utopias and Heterotopias." Architecture/Mouvement/Continuité.
Ginsberg, R. (2004): *The Aesthetics of Ruins*. Amsterdam: Rodopi.
Gros, F. (2014): *A Philosophy of Walking*. London: Verso.
Haraway, D. (2016): *Staying with the Trouble*. Durham: Duke University Press.
Harbison, R. (2015): *Ruins and Fragments: Tales of Loss and Rediscovery*. London: Reaktion Books Ltd.
Heritage Malta (2024): *Malta Biennale.Art 2024 Official Guide*. Malta: Heritage Malta.
Huyssen, A. (2003): *Present Pasts*. Stanford: Stanford University Press.
Ingold, T. (2011): *The Perception of the Environment*. London: Routledge.
Ingold, T./Vergunst, J.L. (2016): "Introduction." In: T. Ingold/J. L. Vergunst (eds.), *Ways of Walking*. London: Routledge, pp. 1-19.
King, H. (2015): *Virtual Memory*. Durham: Duke University Press.
London, B. (2020): *Video Art: The First Fifty Years*. London: Phaidon Press Limited.
Lorraine, T. (2010): "Smooth Space." In: A. Parr (ed.), *The Deleuze Dictionary*. UK: Columbia University Press, pp. 253-254.
Macfarlane, R. (2016): *Landmarks*. UK: Penguin Books.
Massey, D. (2005): *For Space*. London: Sage.
Maxine, A./Nicholas, C.V. (2018): "Olive Oil Production Technology in Roman Malta." In: C. V. Vella/ A. J. Frendo/ H. C. R. Vella (eds.), *The Lure of the Antique*. Leuven: Peeters Publishers, pp. 275-300.
Merleau-Ponty, M. (2004): *The World of Perception*. London: Routledge.
Mitchell, W. J. T. (2002): *Landscape and Power*. Chicago: The University of Chicago Press.
Morgan, D. (2019): "Cave of Darkness – Port of No Return." In H. Iliadou de Subplajo-Suppiej (ed.), *Maleth/Haven/Port Heterotopias of Evocation*. Milan: Mousse Publishing, pp. 94-99.
Pearson, M./Shanks, M. (2001): *Theatre/Archaeology*. London: Routledge.
Rancière, J. (2004): *The Politics of Aesthetics*. London: Bloomsbury Academic.
Rancière, J. (2011): *The Emancipated Spectator*. London: Verso.
Solnit, R. (2017): *A Field Guide to Getting Lost*. Edinburgh: Canongate Books.
Trigg, D. (2012): *The Memory of Place*. Athens: Ohio University Press.

Tsing, A. (2015): *The Mushroom at the End of the World*. Princeton: Princeton University Press.
Tsing, A./Swanson, H./Gan, E/Bubandt, N. (2017): "Introduction. Haunted landscapes of the Anthropocene." In: A. Tsing/H. Swanso/E. Gan/N. Bubandt (eds.), *Arts of Living on a Damaged Planet*. Minneapolis: University of Minnesota Press, pp. 1-14.
Tuan, Y. (1977): *Space and Place*. Minneapolis: University of Minnesota Press.
Vergunst, J. (2016): "Seeing Ruins: Imagined and Visible Landscapes in North-East Scotland." In: M. Janowsk/T. Ingold (eds.), *Imagining Landscapes: Past, Present and Future*. London: Routledge, pp. 19-37.
Xerri, D. "Art That Oscillates Between the Poetic and the Political" (18 May 2024). Retrieved from https://www.maltatoday.com.mt/arts/art/129161/art_that_oscillates_between_the_poetic_and_the_political

Allegorical Ruins and the Possibilities of Human Futures
A Case Study of the Game Lifeless Planet

*Caio Tulio Olimpio Pereira da Costa
and Ana Laura Matos Torquato*

Abstract

This research contextualizes ruins as allegorical constructions that extract a fragment from the flow of History-Destiny projected as a possibility for human futures, viewing life through death and decay, and reflecting not on what was, but on historically unbuilt potentialities. To support this discussion, a qualitative case study is conducted on the independent video game Lifeless Planet (LP), *which involves the exploration of an uninhabited planet with abandoned Soviet-era ruins. The narrative and mechanics of this game are developed as a third-person action-adventure game inspired by Cold War-era sci-fi stories. At its core, it explores and demonstrates issues related to humankind's desire for space travel through both physical and digital, as well as symbolic ruins. Through the immersive experience of interactive digital narratives that highlight possible human futures and aesthetic encounters, the analyses in this research are based on the authors' self-narratives as a methodological approach. This scientific production results from a theoretical, scientific, and essayistic research strongly supported by concepts of immersion, image and imaginary, and experience, considering the ruins in LP as a halo capable of realization, breaking with the idea of the future as conjecture filled with false illusions, and instead as intentions of the future, projects, and perspectives that guide the individual in the present. Therefore, by combining the narrative capacity of games with the evocative power of images and the possibilities of ruins, there are inevitably manifestations of future imaginaries. In this sense, the game serves as a contemporary catalyst for the scientific, environmental, and social materializations and challenges that previously existed only in the collective imagination.*

Keywords

immersion, game studies, experience, ruins, Lifeless Planet

Introduction

The concept of ruins has been a focal point in cultural and historical studies, symbolizing the remnants of civilizations, the passage of time, and the inevitable decay that accompanies human achievement. Ruins, whether architectural, technological, or symbolic, serve as tangible markers of past events and cultural practices, embodying both the grandeur and the fragility of human endeavours. Historically, ruins have been interpreted as sites of memory, where the remnants of what once was evoke contemplation of history, loss, and the transient nature of human existence. As Vella (2010) notes, "To contemplate the ruin, then, is to engage in the play of presence and absence surrounding a fragment that connotes both an original totality and an agent, and process, of ruination" (91). This duality – where ruins represent both a connection to the past and a process of ongoing decay – provides a fertile ground for exploring what has been lost and also what could have been, and by extension, what could still be. The study of ruins thus transcends mere historical inquiry, delving into the realm of potentialities, where the unfinished and the unbuilt offer a lens through which to consider future possibilities.

The interpretation of ruins has evolved to encompass a broader understanding of their role as allegorical constructs. Ruins are no longer solely viewed as remnants of the past but are also seen as symbols that project human fears, aspirations, and the trajectories of civilizations. This shift reflects a growing recognition of the importance of ruins in the collective imagination, where they serve as metaphors for the cyclical nature of history and the potential futures that lie ahead. As Fraser observes, "The ruins are an integral part of the playable landscape, and the kind of play that it produces" (2016: 179). In this sense, ruins are not static objects but dynamic spaces of interaction, where the past, present, and future converge in a complex interplay of meaning and interpretation. This perspective aligns with the cultural and historical understanding of ruins as sites of memory and imagination, where the remnants of the past are imbued with new significance in the contemporary concerns and future aspirations.

The digital age has brought a transformation in how ruins are perceived and engaged with, particularly within the context of video games. In these interactive environments, ruins are not merely aesthetic backdrops, but integral elements of the gameplay experience, shaping both the narrative, fictional/digital universes, and the player's interaction with the game world. Games, as noted by Bogost, "represent processes in the material world – war, urban planning, sports and so forth – and create new possibility spaces for exploring those topics." (2008: 121) This capacity of video games to simulate and reimagine real-world processes extends to their treatment of ruins, which are often depicted as spaces of decay and desolation that can challenge the player to navigate and make sense of the fragmented landscapes. The depiction of ruins in games like *Lifeless Planet* (LP), which will be presented along this research, exemplifies this approach, where the abandoned Soviet-era research stations and desolate environments on another planet serve as both literal and metaphorical obstacles for the player to interact and overcome.

Fuchs states, "if we look at contemporary video games, we find an abundance of ruined buildings, of mould, and of all forms of decay of organic matter and inorganic materials." (2012: 1) These elements of disintegration tend to be relevant to the player's experience and comprehension of the digital environment, encouraging a reflection on the transient nature of human endeavours and potential futures that such ruins represent. The game's use of ruins as a narrative device and a gameplay element, allows players to engage with these themes in a way that could be both potentially immersive and thought-provoking, making the ruins not just a setting, but a key element in the exploration of possible human futures.

The devastation depicted in the ruins that permeate these environments drives the plot forward, symbolizing a literal and figurative obstacle that the player or protagonist must overcome. Through the exploration of these decayed remnants, games engage with themes of loss, renewal, and resilience, allowing players to interact with the physical and emotional weight of what remains and of what must take place.

Analysing ruins in the context of digital narratives like LP offers a unique opportunity to explore the allegorical potential of these spaces as representations of possible futures. Ruins, as they appear in this game, could be more than just the remnants of a failed past; they are projections of potential futures that have been shaped by fictional historical events and human actions. This perspective aligns with Vella's assertion that "Throughout the history of video games, ruins have proven to be a uniquely versatile tool of recapture" (210: 67). The game thus becomes a medium through which the player can explore the intersection of history, memory, and future possibilities, using the ruins as a lens to examine the consequences of human ambition and technological progress.

The relevance of this analysis extends beyond video games, offering insights into broader cultural and philosophical questions about progress, the role of technology in shaping our futures, and the ethical implications of these pursuits. The ruins in video games, like those in the real world, serve as reminders of the fragility of human achievements and the potential for decay that accompanies even the most ambitious endeavours. They also offer a space for imagining alternative futures, where the lessons of the past inform new possibilities. The study of ruins as allegorical constructions in LP can contribute to a deeper understanding of how digital narratives engage with cultural and historical themes, providing a platform for reflection on the potential trajectories of human civilization. This approach enriches the academic discourse on video games and digital narratives and offers a valuable perspective on the role of cultural artefacts in shaping our understanding of the past, present, and future.

Building upon the exploration of ruins, this research is driven by the central inquiry: How do the ruins in LP function as allegorical constructions that project possible human futures? This question arises from the recognition that the ruins are not merely aesthetic or narrative devices but symbols that encapsulate broader existential and philosophical themes. These themes encompass the fragility of human endeavours, the consequences of technological advancement, and the speculative futures these ruins evoke. The game's portrayal of an uninhabited planet, scattered with abandoned research stations and remnants of Soviet-era architecture, invites players to reflect on the cyclical nature of creation and decay, as well as potential trajectories of human civilization. By engaging with these symbolic ruins, the research seeks to uncover how digital narratives explore and critique contemporary issues related to progress and human futures.

Moreover, the ruins in digital narratives are often imbued with political and social significance. They raise questions about the fate of civilizations, the consequences of human's destructive actions, and the potential for regeneration or further ruination. Often, the protagonist is granted the opportunity to not only explore and uncover these ruins, but to attempt to rectify the mistakes of the past, preventing history from repeating itself. This narrative arc – of setting right what was once wrong – offers a sense of catharsis, where the player feels satisfaction in both the restoration of order and the avoidance of future collapse, putting herself in the role of the saviour.

To address this research question, the study sets forth a general objective: To analyze how the ruins depicted in LP function as allegorical constructions that project possible human futures. This goal underscores the intention to explore the symbolic and narrative functions of ruins within the game, examining how they contribute to the construction of speculative futures and challenge notions of progress and destiny. The specific objectives of the research are twofold. First, to conduct a qualitative case study of LP, focusing on the game's depiction of an uninhabited planet with abandoned scientific research stations and Soviet-era ruins. This involves analysing the game's narrative and visual aesthetics to understand how these elements create a cohesive allegorical framework. Second, to investigate how the combination of digital narrative and evocative imagery in LP contributes to constructing future imaginaries and challenges notions of progress and destiny. This objective explores the broader implications of the game's narrative, particularly how it uses the interplay between visual and textual elements to engage players in a reflective process about the future of humanity and the legacies of past ambitions. Through this dual focus, the research aims to contribute to the academic discourse on digital narratives, game studies, and cultural analysis, offering insights into the role of videogames as tools for exploring philosophical and existential questions.

The current research is relevant in today's social, political, and economic context due to its exploration of the symbolic meanings behind ruins and their implications for the future of humanity. In an era marked by rapid technological advancements, environmental degradation, and geopolitical tensions, the concept of ruins as allegorical constructions offers a reflection on the consequences of human actions and the potential trajectories of our civilization. By engaging with these symbolic ruins, the study delves into critical questions about the sustainability of technological and scientific pursuits, the ethical dimensions of space exploration, and the broader socio-political implications of these endeavours. Thus, this research holds significant social relevance, as it encourages a critical examination of contemporary issues through the lens of allegorical ruins, fostering a deeper understanding of the interconnectedness between past, present, and future (figure 1).

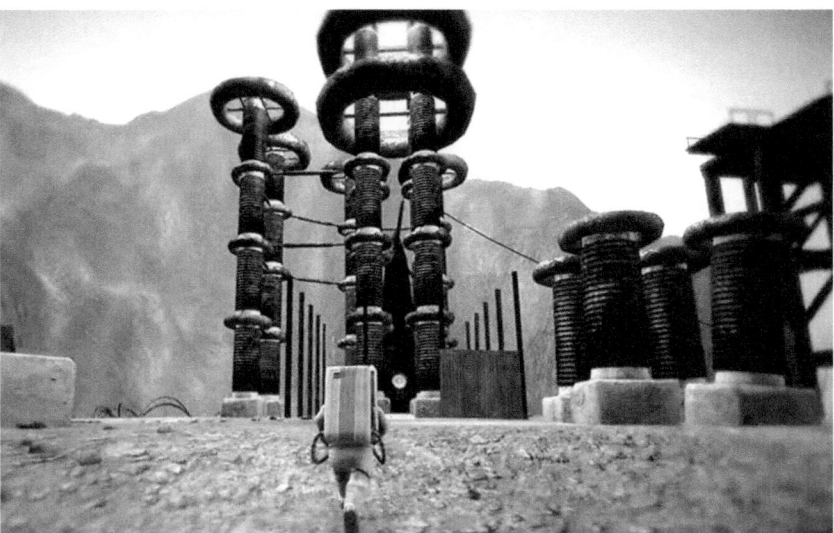

Fig. 1: LP Ruins. Source: LP, 2024.

The case study of LP exemplifies the potential of games to convey complex philosophical and existential themes, challenging traditional notions of narrative and storytelling. By employing a methodological approach grounded in self-narratives/autopoiesis, the research bridges the gap between personal experience and academic inquiry, offering a perspective on how players interact and interpret digital environments. This approach enriches the understanding of the game's narrative and aesthetic dimensions and contributes to the broader discourse on the role of digital media in shaping contemporary cultural imaginaries. Additionally, the study's focus on immersion, image, and the imaginary aligns with current academic debates on the role of visual culture in constructing and deconstructing societal values and beliefs. By situating the analysis within the context of Cold War-era science fiction, the research engages with historical and cultural studies, offering a multidisciplinary perspective that enhances its academic significance.

Theoretical Framework

Brief definition of ruins or the role of ruins on imaginaries of future

The concept of ruins is multifaceted, extending beyond physical remnants to encompass symbolic and historical dimensions. As Vella (2010) notes, ruins can be understood not only as objects but also as processes, events, or outcomes. They represent decay, fall, and the remnants of what once was – whether that be structures, institutions, or even entire civilizations. This complexity is key to understanding ruins as both physical artefacts and as expressions of historical and cultural trajectories. Ruins, therefore, are never static; they embody temporal disjunctions, bridging the past, present, and future.

Roth, Lyons and Merewether (1997) suggest that ruins hold a unique power to evoke the past, reflecting the structures they once represented and the people who constructed, commanded, and utilized them. Yet, in their very state of decay and emptiness, ruins also convey a sense of loss – signifying those who either abandoned, destroyed, or were unable to shield them from the inevitable effects of time. Ginsberg (2004) argues that engaging with a ruin requires, then, active and continuous reinterpretation by the viewer. Navigating a ruin-site involves interpretation on a practical level – determining one's path through disrupted structures – and also on an aesthetic level, as each new perspective reveals a different arrangement of elements, suggesting new ways for these fragments to cohere into a unified form. Through movement, viewers create additional visions of the ruin's inherent structure.

Vella (2010) states ruins embody a complexity that goes beyond their simple depiction as remnants of the past. Macaulay's work *Pleasure of Ruins*, from 1953, highlights that ruins can captivate us, evoking not only interest but a unique aesthetic appeal. Moreover, Woodward's idea that ruins stimulate a dialogue between incompleteness and the imagination sheds light on this allure (Woodward 2002). This interplay of what is present and what is absent captures the essence of a ruins' appeal, drawing the viewer into a space where history's fragments meet the mind's reconstructive impulse.

Furthermore, ruins can evoke a sense of horror and regret, as they represent remnants of a once vibrant past now irreversibly lost, inciting a deep nostalgia for what could or should have been. They serve as symbols of decay and destruction, invoking feelings of sorrow and fear of future devastation. This emotional complexity has made ruins a dominant motif in contemporary media, especially in recent video games where the exploration of ruins plays a critical role in plot and character development, enticing the player to discover and explore the history in these ruined environments. According to Vella (2010), ruins appear with striking frequency as settings in adventure games, exemplified by the archaeological journeys of *Tomb Raider* and *Prince of Persia*. This prevalence reflects the broader resonance of ruin imagery within Western culture, where ruins are prominent across various media. The association of games with genres like fantasy, sci-fi, and Gothic horror – genres that frequently incorporate ruins as a symbolic element – further reinforces the prominence of ruins in digital game landscapes.

Building on this understanding, Lombardi (2024) suggests that traces found in ruins represent a fragmented historical experience, marked by unfinished elements, temporal discontinuities, and layers of meaning that transcend singular timelines. These traces do not merely point to a completed past; instead, they reflect an interplay among past, present, and future, resonating with Benjaminian ideas where ruins hold enduring marks that resist erasure. Ruins, in this light, partially open events, recovering spatial and temporal fragments that offer on-going connections to various moments in time. In discussing the "poetics of ruin", Lombardi (2024) also emphasizes that these remnants invite reflection through their visible, enduring traces. They stimulate our imaginations, prompting a continuous reassignment of meanings to what remains. This openness allows us to link objects to events, interpret their elements, and weave connections across past, present, and potential futures.

Lombardi (2024) questions whether images serve as ways to recall the past or as frameworks to shape our imagined interpretations of it. She notes that, consciously or unconsciously, we interpret images based on idealized visions of the past, not simply stating something about the images but bringing forth our projected meanings. The interpretation of images, then, holds a "poetic responsibility," which includes the ethical creation of images tied to memory and imagination. Finally, the author asserts that images are inseparably linked to ruin, bearing traces of a past that challenges us and enriches our understanding. While we cannot fully grasp the meanings and uses images held centuries or millennia ago, they nevertheless provoke a form of "inter-humanity," where the past and present dialogue, sparking reflections that connect us across time.

Ruins, whether composed of stone, wood, or steel, evoke a strong collective reaction, drawing attention to the intersection of past, present and future. This intersection, where time collapses into a shared moment, is particularly prominent in video games, as they begin from a place of disorder. Unlike other media, where ruins often symbolize destruction or the past left behind, video games use this aesthetic as a starting point for interaction and transformation.

In video games, players are frequently tasked with restoring order from chaos, a fundamental mechanic in many titles. For example, even abstract games like *Tetris* require players to create structure from the falling, chaotic blocks by aligning them into rows, which are then erased, reordering the space and restoring a form of order (Chandler 2014) In this sense, players take on the role of a "restorer," someone who actively changes the world through their actions, mirroring the core idea of engaging with a broken or ruined environment. The environment becomes an aesthetic backdrop that reflects the central gameplay goal of restoring balance or understanding the context of a fractured world.

Games like the *Fallout* franchise present players with the opportunity to mend or deepen the fractures of a broken world. In contrast, the *Dark Souls* franchise offers a setting where the player's actions either alleviate or worsen the despair of a godless, abandoned land, represented by decaying ruins filled with the restless dead (Chandler 2014). Similarly, *Bastion* takes place after a cataclysmic event, where the player reshapes a broken world, walking across the void and reforming it piece by piece. These examples illustrate how video games immerse players in worlds where ruins are not merely objects to observe but environments to engage with, reshape, or repair. The presence of ruins in video games challenges players to confront the fragility of systems they inhabit. Decaying buildings or abandoned shrines in these games remind that social orders and structures can, just like the ruins, collapse or disappear. This interaction with ruins through gameplay encourages reflection on the transient nature of power, society, and civilization.

The aesthetics of ruins in video games transcends passive observation. It becomes a transformative experience. While traditional representations of ruins often encourage detached reflection on the fall of past civilizations, video games offer the player an active role in engaging with these remnants. The decay of former civilizations in games is not just a backdrop but an integral part of gameplay, where the past is reconfigured through interactive mechanics. Players no longer merely observe ruins as historical curiosities; instead, they actively engage with them, reshaping the past to fit new purposes within the game's world. This shift from passive observation to active engagement reflects the subversive potential of the ruin aesthetic, offering a dynamic and hands-on relationship with history and memory, allowing the player to shape the past beyond traditional media.

In essence, video games not only allow players to witness the decay of civilizations but invite them to intervene, to remake the past, project the future, and to discover how it can be transformed. The ruins in games symbolize not just the passage of time and the fall of societies, but also the potential for renewal, even if only in the digital environment. Through this engagement, players are reminded that the forces of time and decay can be momentarily halted or altered, if only through the digital manipulation of the past.

In many games, ruins serve not merely as a backdrop, but as a pivotal element in the progression of the narrative. The act of investigating and traversing these destroyed landscapes often becomes a metaphor for confronting the past. As players navigate through devastated worlds, they are encouraged to uncover the history behind the destruction, piecing together fragments of forgotten societies, political structures, and human experiences that once flourished within these spaces.

The game world, then, can be understood as a layered repository, where the remnants of past events are embedded within its environment, allowing it to function as a complex space rich in historical traces (Vella 2010). This concept extends beyond mere depictions of ruins; it suggests that the game world itself serves as a semiotic structure, enabling players to interpret meaning through its layered representations of time and decay.

In post-apocalyptic games this interpretative depth is intensified by the imagined histories they propose, where catastrophic events reduce civilizations to ruins, casting destruction as an almost natural and expected outcome (Fraser 2016). This approach romanticizes ruin, transforming it into a symbol of heroic loss rather than a reminder of ordinary devastation. The combination of hypothetical histories and the aesthetic of decay provides players with a unique experiential framework, one that invites them to explore ruins not just as remnants of the past, but as dynamic symbols that enrich the narrative space of the game.

Ruins can be considered fragments of History or Fictional History suspended in time, preserved as eternal symbols of hope while simultaneously reminding us of human mortality and the inherent fragility of our existence. The German word "Ruinenlust", which can be translated to a fascination or love for ruins and destruction, encapsulates the paradoxical allure these remnants can hold. According to Simmel (1958), this fascination culminates in a return to the nature of inorganic things, a return to the spirit through the inescapable and re-signifying force of nature. The triumph over the human will, that was present there in wood, marble, cement and lime, and which leaves something new, as if the saying "for dust you are and to dust you shall return" were, in fact, an inescapable promise.

The sight of ruins also brings with it the eminence and fear of witnessing the fall of empires and civilizations. The visual and emotional impact of such images provokes an awareness of both the grandeur of what once was, and the vulnerability that accompanies power. In these contexts, victory becomes transformative, not merely a triumph over immediate challenges, but an emotional and intellectual reckoning with the potential for ruin in the future. The power of ruins lies in their dual capacity to inspire awe and fear, reminding us of the cyclical nature of rise and fall and urging us to heed the lessons of history contained in these fragments.

Regarding the haunting nature of ruins and what they can represent and contain, Didi-Huberman (2008) talks about the ruins of Auschwitz and the destruction in the concentration camps. Although they no longer exist, these ruins still contain vivid memories of horror, symbolizing the death of an ideology and the desire not to repeat the mistakes of the past. Didi-Huberman (2008) notes that these ruins are slowly being forgotten, as they continue to be categorized as distant memories of unknown monsters, long buried and destroyed, while new executioners emerge and the nation pretends not to hear their endless cries.

The ruined visuals that are increasingly present in video games bring out stories and desires, transforming the ruins into a threshold between the visible and the invisible. According to Didi-Huberman (2008), it is up to images to reveal sensitive points, their dialectical unfoldings, memories, desires and conflicts contained therein. To become sensitive is to go through this dialectic, to see what is not explicit in the images, bringing out feelings and memories. Images of ruins are powerful for interpreting our own moment, offering a window into human fragility and the unstoppable flow of time. The history of images, for Didi-Huberman, can be told as "an effort to visually transcend the trivial contrasts between the visible and the invisible" (2008: 133). Ruins contain fragments of time and spirit, transported into an object that fuses past and present, nature and spirit (Simmel 1958); This tension reminds us that decadence is an unstoppable, egalitarian force, lacking discernment about what lies before it, transforming and reinterpreting matter into a deeply alive and ever-changing expression.

Immersion and Experience on Game Studies as Epistemological Anchor

The concept of immersion encompasses a wide range of interpretations and applications, varying according to academic context or common understanding. Machado (2014) notes that the term has multiple interpretations and distinct associations and applications, referring to the act of diving or deeply engaging with something, a concept widely explored by various authors.

The idea of immersion, described as a dive into the ocean by Murray (2003), can lead to a possible connection to Huizinga's (2014) concept of the Magic Circle. From a psychologically immersive experience, we seek a sensation akin to diving into the ocean or a swimming pool: the feeling of being enveloped by a distinct reality, as contrasting as water is to air, one that fully captures our attention and engages our entire sensory system (Murray, 2003).

In the context of ruins and video games, the individual is usually led to experience an engaging story and ambient that promotes a temporary disconnection from reality. Such an immersive experience is compared to the "siren's song" which can draw the viewer into new realities of decay. According to Murray, our minds are particularly "programmed to tune into stories with such intensity that it is possible even to obliterate the world around us" (2003: 101), a phenomenon heightened by the Magic Circle's capacity to create engaging and convincing realities.

For Gadamer (1997), in his propositions on aesthetic experience and hermeneutics itself, there is no direct mention of a concept of immersion, although the author uses this word to highlight the possibility of a form of emotional inhabitation within the media experience, where the individual not only observes but participates in and complements the work. In this sense, engaging with a narrative provides a profound sense of belonging, becoming an experience that reaches all the senses of the participant. Immersion, functioning as the Magic Circle's "siren song" transforms the observer into an active and emotional participant who recreates and reinterprets the experienced content.

The phantasmagoria of the presence of the ruins in video games is not only reflected in the obsession with novelty and utopian ideals of capitalist mass production, but also in teleological histories that envision futures as realized, casting history as a progression into an untouched, unrealized time, and thereby concealing the world's true state of catastrophe and decay. According to Fraser (2016), in games that depict such a "fulfilled" future through catastrophic collapse, ruins emerge to unsettle the immerse player, peeling back illusions of safety and progress. These ruins simultaneously affirm and challenge the view of history as a cycle of continuous novelty, beginnings, and endings. It is the ruin – invoked for play beyond the immediate catastrophe – that enables this resistance, exposing the phantasmagoric illusions of modernity itself.

These metaphors reinforce that immersion is not limited to the physical realm. The "siren's song" suggests an involvement that transcends mere displacement between realities. For Machado (2014), this process is not simply a "leap" from daily circumstances to the pages of a book or the screen of a cinematic narrative, but rather implies complete immersion, where there is a genuine shift in reality and medium. In the perspective of digital games, the Magic Circle and immersion stand out for their ability to simulate cognitive and narrative processes through images and technology. Like cinema and other media, contemporary digital games exert a significant suggestive power on the mind, using movements and effects that enhance narrative and visual potential. In this context, we are witnessing an era marked by the rise of imaging technologies, which, as Grau argues, produce "an encompassing sensory and visual sphere similar to life" (2007: 15), consolidating games as an immersive experience space that rivals and/or builds reality itself.

The video game reproduces essential mental processes through images. While each thought is narratively constructed – whether by telling a story, recalling events, planning the day, or dreaming – it often relies on images as its foundation. Games, like cinema, theatre, animations, television series, and soap operas, when representing ruins, create a visual language with significant narrative potential for thought. This language incorporates cinematic movements and effects, contributing to the construction of ideas.

Ruined landscapes and structures, left as abandoned spaces, convey a deep sense of loss. According to Merewether (1997), these ruins offer a glimpse of a past once vibrant, representing an origin that can no longer be reclaimed. Their existence signals an irrevocable absence, reminding us of our finitude, where disruption and continuity coexist, emphasizing the need to persist amid decay. Experiences and thoughts are deepened through an immersive process, intertwining reflection and presence to navigate the coexistence of disruption and continuity.

Brockmeier and Harré (2003) present arguments that align with Murray's (2003) ideas, reinforcing the notion that narrative generates multiple possibilities for interaction, consequently, experiences and immersion. They argue that narrative functions as an effective intermediary model between individual experience and interpretation of the surrounding world, enabling a clear and acceptable understanding. As Murray states, "narrative is one of our primary cognitive mechanisms for understanding the world" (2003: 10). This can be perceived, for instance, in video games such as:

a) *The Last of Us* (Naughty Dog 2013), which presents a post-apocalyptic world where nature reclaims urban and suburban ruins. Abandoned buildings, crumbling skyscrapers, and overgrown streets contribute to an immersive sense of humanity's fragility, pushing players to imagine the lives once lived in these desolate spaces. The ruins here evoke both tension and beauty, enveloping players in a landscape where every ruined corner tells a story.

b) *Shadow of the Colossus* (Team Ico, 2005), which the game's open-world design immerses players in a desolate landscape filled with ancient ruins, remnants of a long-lost civilization. These ruins add layers of mystery and awe, creating a haunting atmosphere that deepens as players uncover the story. The emptiness amplifies the feeling of isolation and scale, as players must navigate vast, silent expanses to confront each colossal opponent.

c) *Horizon Zero Dawn* (Guerrilla Games 2017), which players explore a world filled with the ruins of an advanced civilization, now overtaken by nature and robotic creatures. These ruins serve as visual and narrative tools that immerse players in a storyline cantered on the collapse of modernity. As players uncover the remnants of this lost society, the ruins build a compelling mystery about what caused the downfall and how it affects the world's current inhabitants.

In each of these, ruins are not simply background elements; they are integral to the narrative and emotional experience, enhancing immersion by evoking histories and atmospheres that drive players to explore, interpret, and engage with the world in a more profound way. The decay and history present in these environments encourage a reflective, immersive experience that prompts players to ponder the fate of civilizations, both fictional and real.

This theoretical framework highlights the critical role of narrative and ruins in shaping immersive experiences in video games, reinforcing the connection between storytelling, player engagement, and cognitive processes underpinning immersion. By exploring the interplay between narrative, experience, and immersion, we can better understand how games create compelling worlds that resonate with players on multiple levels. This sets the foundation for examining how ruins, as narrative and environmental elements, contribute to the immersive quality of video games, offering both a symbol of loss and a space for exploration.

Self-Narratives as Methodology: The Researcher-Player

The research employs self-narratives, also referred as autoethnography (Santos 2017) or Autopoiesis processes (Maturana/Varela 2001), as a qualitative research method to explore the multifaceted interactions with LP. From an overarching perspective, this analytical tool aims to provide greater consistency to the data collected and facilitates the analysis of possibilities in more expressive immersive experiences in games. With this approach, it is possible to examine how experiences in meticulously simulated realities, integrating the real and the virtual, affect both the researcher-player and the feedback derived from the interaction between the environment-organism and the game-player (Dewey 2010). This method enables different comparisons between the analyses carried out, based on a dual first-person perspective.

In a brief summary, Self-Narrative is a method grounded in a "triadic model", based on three main orientations: the first is a methodological orientation, rooted in ethnography and analysis; the second, a cultural orientation, which involves the interpretation of: a) lived experiences (through memory/subjectivities), b) the relationship between the researchers and the subjects (and objects) of the research, and c) the social phenomena being investigated; and the third, the content orientation, which is based on autobiography with a reflective character (Santos 2017).

This context emphasizes that reflection is an inherent and inseparable step of the investigation through Self-Narratives, as this process organically generates a constant awareness, evaluation, and reassessment made by the authors based on their own contributions, influences, research methods, and the results generated by the method (Santos 2017).

There is a crucial point regarding the Self-Narratives: The structured method and its procedures go beyond what this journey can encompass or limit. According to Whyte's (2005) work on this methodology, the researcher is often an external individual in relation to the research object, not necessarily knowing where the process will lead, and the researched territory may also be new to their reality. At the same time, in order for the external interference of the environment to be acknowledged, the researchers, as external subjects, must constantly affirm this role, which usually places them in a state of alert that prevents full immersion into the object/context of the research. However, in the context we propose, this interference is an inherent part of the process and inseparable from this construction.

This unconventional approach presents a specific and unique authenticity, offering an analysis of what was or was not mobilized by video games based on the experience of living technological narratives through immersion in investigative organisms. Ferreira (2014) argues that Self-Narratives are a practice of personal development that can lead to intellectual growth and broaden one's perspectives on the world around us. Furthermore, as a process of (self)formation, it contributes to expressing intense movements of everyday life, spontaneity, and even temporality, forming a tripod that gives meaning to a lived life with the potential for sharing and exchange (Silva/Rios 2018).

An autobiographical approach is embraced within the field of ruins and videogames, as it is nourished by what is constructed in the life of the subject who narrates. This subject, by valuing the "process of formation and the meanings constructed through trajectories of formation", whether formal or informal, contributes to the understanding of the subject in development (Silva/Rios 2018: 59). Thus, the qualitative approach is essential in this flexibility, as it provides experiential intimacy. In this way, the present research is grounded in this perspective. The methodological foundations described in this section are therefore theoretical and reflective in nature. Consequently, the material collected and discussed is evaluated through various interactions, capturing a broad range of information.

The qualitative research method adopted in this work is influenced by St. Pierre's (2018) critiques of the disconnect between ontological aspects in research. The author argues that methodology should not be separated from epistemology and ontology, as doing so risks reducing it to a mechanized process, confined to methods and techniques. She emphasizes that it has become clear that the humanistic qualitative approach could never make sense without proper consideration of ontology, representing a methodology that was compromised from the start, although this realization was not fully apparent at the time (St. Pierre 2018).

Autobiographical research encounters its own specific challenges, not merely serving as a "first-person diary of experiences". According to Mello, Murphy and Clandinin (2016), narrative investigation, another way to refer to Self-Narratives, requires continuous, self-reflective work that encourages us to examine who we are in the world and how we shape the world in which we live. This process of self-confrontation is not easy; it requires a careful and critical examination of our own identity and the ways in which we influence our reality.

These authors also provide a definition of narrative research for this method, understanding it as the study of experience as it is lived and narrated. According to Mello, Murphy and Clandinin (2016), narrative research is more than just telling or living stories. While its main characteristic is the study of experiences as they are lived and shared, narrative research goes beyond mere storytelling, involving a deep reflection on the meaning of these experiences and the manner in which they are conveyed.

The study of ruins in video games requires a dual focus on the immersive experiences of researcher-players and the virtual environments they interact with. Through the lens of immersion and experience, researchers explore the symbolic and emotional impact of ruins, as these environments trigger both reflective and embodied responses. This method allows for an in-depth analysis of the interplay between narrative, player agency, and ruinous landscapes, facilitating a deeper understanding of how these virtual spaces evoke feelings of loss, continuity, and the passage of time.

Lifeless Planet – "It's a small world after all, comrades"

LP is a sci-fi video game developed by independent American studio Stage 2 released in 2014. It combines action, adventure, and puzzle-solving, with an emphasis on exploration and survival. Set on a barren alien world, the game follows an astronaut who, after crash-landing, navigates desolate landscapes with limited resources, particularly oxygen, which constrains travel distances and heightens immersion. As the player ventures deeper, they discover a strange human connection, compelling further investigation into the remnants of human presence in abandoned Soviet landscapes.

The protagonist's primary tools are a jetpack with limited fuel and a robotic arm, both essential for overcoming obstacles and uncovering hidden elements. LP's narrative centres on discovery and human resilience. Soviet-era outposts reveal traces of human presence in this lifeless landscape. The story unfolds through environmental cues, notes, and decayed artefacts, which echo past human hopes and failures. This minimalist narrative approach invites players to interpret the story and derive meaning from fragmented information, creating space for reflective engagement.

The ruins encountered in LP are integral to the game's thematic and aesthetic composition, serving as both physical and allegorical elements. These decaying Soviet-era research stations, as the player discovers, evoke a melancholic atmosphere, embodying what Simmel (1958) describes as the process through which nature reclaims and transforms human constructs. These structures, once symbols of Soviet ambition and scientific hope, now stand as silent witnesses to both human resilience and vulnerability. Through these ruins, the game suggests the inevitability of decline and the ephemeral nature of human achievement, juxtaposed with the vastness of the alien landscape. This portrayal of ruins echoes real-world historical sites and abandoned projects, drawing parallels between the in-game world and our collective past. As players interact with these structures, the game evokes a reflection on civilization's cyclical rise and fall, encouraging them to ponder what remains once human efforts fade (figure 2).

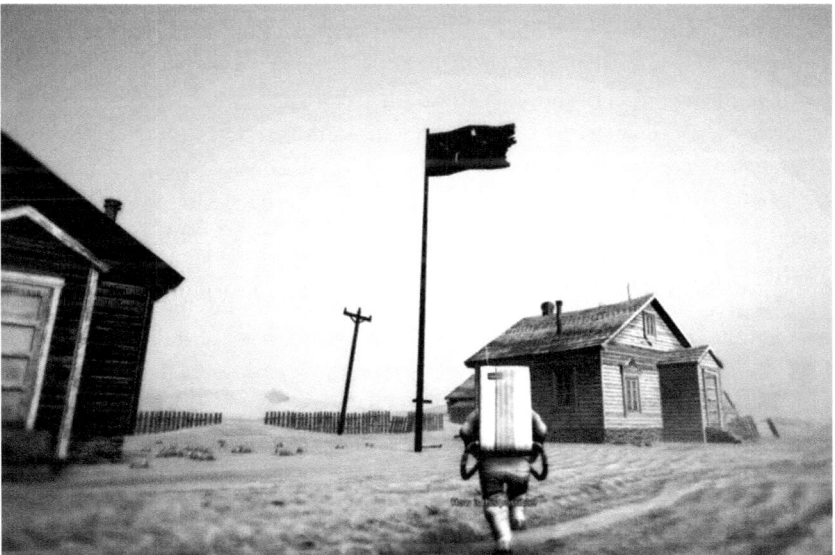

Fig. 2: Soviet Ruins. Source: LP, 2024.

As players navigate LP, they not only engage with its environment but also actively interpret and reconstruct its story. This immersive experience positions players as both participants and observers, blending self-reflection with exploration. Through the lens of self-narrative, players become researchers in their own right, weaving personal experiences and reflections into their understanding of the game's themes. The limitations and mechanics within – such as scarce oxygen and the expansive, uninhabitable landscape – amplify the player's sense of isolation and vulnerability, provoking introspection on survival and resilience. This interaction aligns with Dewey's (2010) concept of experience, wherein the environment becomes a space of mutual influence and reflection. By navigating the game's environment and interpreting its narrative elements, players experience the process of knowledge construction, gaining new insights into the transience of human endeavours.

LP serves as a poignant medium for contemplating future possibilities through the symbolism of ruins. These desolate structures, relics of a bygone era, compel players to reflect on human ambition, resilience, and the relentless pursuit of knowledge. The game poses questions about the sustainability of progress and the legacies we leave behind, especially in the face of environmental and existential challenges. By confronting players with the remnants of a failed utopian project on an alien world, LP invites speculation on the trajectories of civilization and the consequences of unchecked ambition. These ruins become a reflective mirror, offering players an opportunity to project personal and collective fears, hopes, and questions onto an imagined future. The game thus fosters a discourse on the cyclical nature of progress, loss, and potential rebirth, prompting players to consider what "future" truly means in the context of human existence and the environmental footprints we leave behind.

The Soviet ruins occupy a haunting, surreal space within the barren alien landscape. These structures, scattered across desolate terrain, are artefacts of a long-abandoned human mission. Visually, they consist of derelict research stations, broken machinery, crumbling bunkers, and scattered remnants of scientific paraphernalia. Each ruin is rendered with stark, minimalist detail, its weathered surfaces and decayed walls showing clear signs of the passage of time.

These ruins tell a silent story of human ambition – the remnants of Soviet exploration and technological prowess that have now been absorbed by the alien environment. The disintegration of these structures represents the breakdown of once-grand human projects, echoing a theme often found in science fiction: the eventual decay of even the most advanced civilizations.

Soviet symbols represent a bygone era of geopolitical competition and scientific ambition, marking the period when the Soviet Union invested heavily in space exploration and technology. In LP, these Soviet ruins become markers of both this ambition and its limits, symbolizing the drive to conquer the unknown at any cost. Within the game's fictional context, they are evidence of human aspirations projected onto the cosmos, suggesting a narrative where the Soviet Union's fervour for expansion led them not only into space but to colonize and research other planets.

The purpose behind these ruins, as hinted by environmental clues within the game, was to set up a sustainable presence on this alien planet. The complex of research labs, habitation modules, and makeshift facilities reflects an organized yet desperate attempt at survival and scientific study. This portrayal evokes real-world Cold War-era experiments and the isolated research stations that defined the Soviet Union's presence in remote locations on Earth. However, the fact that these ruins are now abandoned and in disrepair paints a stark picture of human fallibility, especially in the face of unknown environments.

The ruins engage us, as player-researchers, emotionally, drawing us into a contemplative space. The aesthetic experience moves beyond mere visuals to tap into existential themes – questioning the sustainability of human ambition, the ephemerality of achievements, and the inevitable decline that awaits all human endeavours. These structures become symbols of humanity's tendency to impose its will on unfamiliar worlds, often without considering the broader consequences. By allowing players to piece together the story through exploration, the game invites an active engagement with these allegories, encouraging players to confront the ethical and philosophical implications of exploration and technological ambition.

Vella states there are ruin-site and ruin-situation. These coexist in LP. The first, "refers to the ruin purely as a physical entity, that is, an aggregation of architectural forms a fragment arranged in a specific topographical locus" (2010: 4), while the other considers both the original site as well as the process of ruination. This emotional response is amplified by the game's use of ambient sounds – howling wind, creaking metal, and distant echoes – which create an oppressive sense of isolation that is also part of the ruination. These auditory cues heighten the immersion, reinforcing ruins as spaces where players can experience connections to the themes of loss and impermanence. As players engage with these ruins, they experience an "aesthetic arrest," a moment where the eerie beauty of decay and the weight of human history intersect to create a deeply introspective experience. The ruins become a mirror through which players can examine their own perceptions of progress and the potential consequences of relentless human ambition. This experience-immersion encourages players to step outside themselves, to reflect on both the grandeur and the limits of human endeavours, and to consider the legacy that their own actions leave behind.

Conclusions

This research contributes to game studies, digital narratives, and visual analysis by examining how video game ruins can reflect on human history and envision future possibilities. It enhances both academic knowledge and the on-going discourse on digital culture and society.

This analysis highlights how LP uses ruins as narrative and experiential elements, positioning them as catalysts for reflection on human ambition and future potential. Each pillar aligns with a qualitative case study approach, inviting player and researcher to engage with the game's symbolic layers. In LP, the ruins are not mere remnants of a failed colonization, but symbolic of broader themes like ambition, technological progress, and their consequences.

The imagery of ruins in LP emphasizes desolate landscapes as artefacts that bridge past-present-future. This underscores images as connectors between the visible-invisible, where the gaze upon absence embodies subjective identification (Mondzain 2009).

By embedding ruins within a Cold War-inspired narrative, LP invites players to consider historical legacies' impact on future trajectories. The Soviet remnants represent civilization's unsustainable concept of "advancement" over nature.

This cycle reveals the connection between human structures-nature, underscoring that all constructions, regardless of advancement, are fated to succumb to natural forces. It serves as both a cautionary tale about humanity's environmental impact and a reflection on human endeavour's transience, reminding us that nature reclaims all, blending humanity's legacy back into its origins.

References

Bogost, I. (2008): "The rhetoric of video games." In K. Salen (Ed.), The ecology of games (pp. 117-140). MIT Press.

Brockmeier, J./Harré, R. (2003): "Narrativa: Problemas e promessas de um paradigma alternativo." Psicologia, 16/3, pp. 525-535.

Chandler, D. (2014): Video games and the aesthetic of ruins. Kill Screen Portal. (killscreen.com/previously/articles/videogames-and-aesthetic-ruins/). Last Visit: Jan. 28, 2025.

Dewey, J. (2010): "A arte como experiência." Martins Fontes.

Didi-Huberman, G. (2008): "Images in spite of all." University of Chicago Press.

Ferreira, L. (2014): "Professores da zona rural em início de carreira: Narrativas de si e desenvolvimento profissional." (Doctoral thesis). UFSCar.

Fraser, E. (2016): "Awakening in ruins: The virtual spectacle of the end of the city in video games." Journal of Gaming & Virtual Worlds, 8/2, pp. 143-160.

Fuchs, M. (2012): "Ruinensehnsucht: Longing for decay in computer games" [Conference presentation]. Apocalypse: Imagining The End, Mansfield College, Oxford.
Gadamer, H.-G. (1997): "Verdade e método." Vozes.
Ginsberg, R. (2004): "The aesthetics of ruins." Rodopi.
Grau, O. (2007): "Arte virtual." Senac.
Huizinga, J. (2014 [1938]): "Homo ludens." Perspectiva.
Lombardi, K. (2024): "Poéticas da ruína: Zones of exclusion." Acta.
Machado, L. (2014): "Da caverna ao Holodeck" [Conference presentation]. VIII ABCiber, São Paulo.
Maturana, H./Varela, F. (2001): "A árvore do conhecimento." Palas Athena.
Mello, D./Murphy, S./Clandinin, J. (2016): "Introduzindo a investigação narrativa nos contextos de nossas vidas." Revista Brasileira de Pesquisa (Auto) Biográfica, 1/3, pp. 565-583.
Merewether, C. (1997): "Traces of loss." In M. Roth et al. (Eds.), Irresistible decay: Ruins reclaimed (pp. 25-40). Getty Research Institute.
Mondzain, M. (2009): "A imagem pode matar?" Vega.
Murray, J. H. (2003): "Hamlet no Holodeck." Itaú Cultural: Unesp.
Roth, M./Lyons, C./Merewether, C. (1997): "Irresistible decay: Ruins reclaimed." Getty Research Institute.
Santos, A. (2017): "O método da autoetnografia na pesquisa sociológica." Plural, 24/11, pp. 54-70.
Silva, F./Rios, J. (2018): "Narrativas de si na iniciação à docência." Educação e Formação, 3/8, pp. 40-60.
Simmel, G. (1958): "Two essays." The Hudson Review, 11/3, pp. 371-385.
St. Pierre, E. (2018): "Uma história breve e pessoal da pesquisa pós-qualitativa." Práxis Educativa, 13/3, pp. 1044-1064.
Vella, D. (2010): "Virtually in ruins: The imagery and spaces of ruin in digital games" (Master's dissertation). University of Malta.
Whyte, W. (2005): "Sociedade de esquina." Zahar.
Woodward, C. (2002): "In Ruins". London: Vintage.

The Ruins Of Home, Hearth, Kingdom, and Man in Breaking Bad
Ruins, Ruination and Ruin Porn in 'The ABQ'

Michael J.T. Stock

Abstract

Working from a transdisciplinary approach that combines cinema and media studies with architecture and cultural geography, this essay analyzes the pivotal role of ruins in the influential television program Breaking Bad (2008-2013). Utilizing a theoretical framework that incorporates John Ruskin's concept of ruins as noble truth, Georg Simmel's view of ruins as uneasy collaboration of humanity and nature, and Tim Edensor's recognition of the ruin as an essential element of modernity, I examine the array of ruins that populate Breaking Bad – from the ruins of Walter White's (Bryan Cranston) family home to the ruins of Albuquerque that comprise creator Vince Gilligan's vision of the city as "The ABQ" which provides the setting for the show.

Keywords

Breaking Bad, Better Call Saul, *ruins, ruin porn, phenomenology, cultural geography, media studies*

Whether we like it or not, modern ways are going to alter and in part destroy traditional customs and values.

WERNER HEISENBERG (1958)

Do not let us deceive ourselves in this important matter: it is *impossible*, as impossible to raise the dead, to restore anything that has ever been great or beautiful in architecture.

JOHN RUSKIN (1849)

Ruin porn is the new sublime.

SIOBHAN LYONS (2018)

What is left of hearth and home when the rot sets in and the squatters take over? This is the question that both the character of Walter White and viewers of *Breaking Bad* are confronted with in "Blood Money," the ninth episode of the fifth and final season of the television show which originally aired on August 11, 2013 – written by Peter Gould and directed by Bryan Cranston (the actor who plays Walter White on the show). The episode opens in eerie flash-forward □ or, □prolepsis,□ as Gérard Genette would refer to it (1983) □ to an undetermined point in future-time when Walter White revisits his now-abandoned house after his identity as the notorious drug czar "Heisenberg" has been revealed. Here we see the family home transformed to ruins – the living room littered with dirt and trash, and marked by graffiti paying tribute to the fallen kingdom of Heisenberg in the form of his name spelled out in massive capital letters that stretch across the entirety of one wood-paneled wall. The new inhabitants of the ruins – teenagers skateboarding in the now drained and grafitti'd swimming pool in the back yard – are inheritors of the new world of Heisenberg, built on the foundations of his unparalleled blue meth.

For Walter, the ruins of his family home mark the ruination of both family and empire. This revelation of traumatic modernity is a vision of the Ruin as wound. It is simultaneously the Ruin of Modernity, Ruin as remnant, Ruin as memorial, Ruin as mnemonic, and certainly the Ruin of Walter White. But at the same time, it is also the Ruin as process, Ruin as knowledge, Ruin as politics, Ruin as ideology, and arguably even the Ruin as resistance – all seen within a tradition of the Ruin that characterizes Vince Gilligan's vision of Albuquerque, New Mexico in the Southwestern part of the United States as a city strewn with ruins -- simultaneously physical, mental, cultural, economic, and moral.

Working from a transdisciplinary approach that combines cinema and media studies with architectural theory and cultural geography, this essay examines the pivotal role of ruins in *Breaking Bad* in this key sequence. I will be utilizing a theoretical framework that spans three centuries, incorporating John Ruskin's concept of ruins as noble truth in the 19th century, Georg Simmel's view of ruins as uneasy collaboration of humanity and nature in the 20th, and Tim Edensor's recognition of the ruin as an essential element of modernity in the 21st. This essay is the first part of a larger research project that examines the array of ruins that populate *Breaking Bad* and *Better Call Saul* (the equally popular prequel series that ran from 2015 to 2022) – from the ruins of Walter White's family home to the ruins of Albuquerque that comprise creator Vince Gilligan's vision of the city as "The ABQ" which provides the setting for both shows.

The house as home, the poetics of space

For the character of Walter White, the house on 308 Negra Arroyo Lane was the first and only home he ever owned – a family home shared with his wife, Skyler, his teenage son, Walter Jr., and eventually his infant daughter, Holly. (This is the fictional address used on the show, as voiced by Walter in the first lines of dialogue on the pilot episode. In actuality, the house is located at 3828 Piermont Drive NE). In the opening sequence of "Full Measure," the thirteenth (and final) episode of the third season (06/13/10), we join the young White couple in a flashback to sixteen years earlier when they first looked at the house. While Skyler seems convinced it is a perfect place to build a family, Walter worries they are setting their sights too low, but eventually gives in, reluctantly calling it their 'starter home' and insisting they won't be there forever. "Why buy a starter house when we'll have to move out in a year or two?" Walter asks Skyler. "Why be cautious? We've got nowhere to go but up."

This sequence effectively serves as a bookend accompaniment to "Blood Money," marking Walter's first and last visit to his family home respectively. While "Full Measure" presents the house in its raw state, "'pregnant' with possibility", as Angelo Restivo describes it, the ruins of the home revealed in the later "Blood Money" reveal the failure of Walter as husband, father, educator, and drug kingpin (Restivo 2019: 57). While still 'pregnant with possibility,' those options are now accessible only to the teenage squatters, who have realized new possibilities of the house and yard – now a party house, skate park, and refuge from both the law and mainstream society. The chain link fence set up outside the house by police ostensibly to keep people out just as surely serves as a protective mechanism and semi-permeable membrane, simultaneously severing it from the neighborhood; a house now marked both by and for crime. The fact that Walter is so easily able to slip through the fence on a practical level reveals the street smarts he has acquired over the course of the show, a rite of passage that allows him entry more assuredly than the fact that the house used to be his.

Even as a house marked by and for crime, to Walter, it is home. That is, it still *means* home to him (even if the city of Albuquerque has declared that it is no longer his). For Walter, as for phenomenologist Gaston Bachelard, "our house is our corner of the world... our first universe, a real cosmos in every sense of the word" (Bachelard 1994 [1964]): 4). For both men, the house that is a home is the preeminent safe space that both embodies dreams and enables daydreaming, prompting Bachelard to declare in *The Poetics of Space* in 1964 that the "house image would appear to have become the topography of our intimate being" (ibid: xxxii). As Bachelard sets out in the introduction to his influential book:

"In the course of this work, we shall see that the imagination functions in this direction whenever the human being has found the slightest shelter: we shall see the imagination build 'walls' of impalpable shadows, comfort itself with the illusion of protection—or, just the contrary, tremble behind thick walls, mistrust the staunchest ramparts. In short, in the most interminable of dialectics, the sheltered being gives perceptible limits to his shelter. He experiences the house in its reality and in its virtuality, by means of thought and dreams." (ibid: 4)

For Walter White the family home is initially meant to serve as a foundational base for building both his family and his career – even if he foresees it in the beginning as limiting, with not enough rooms to accommodate the grandeur and scale of his vision. The notion of it as a space for protection is both taken as a given and taken for granted – until his final return to the house which is now no longer his own and easily breached. Still, it remains the family home, or, as Bachelard would describe, an "oneiric house, a house of dream-memory, that is lost in the shadow of a beyond of the real past" (ibid: 15). For Walter White, as for Bachelard, there is only one oneiric house which marks him, secretly storing pathways that allow access to his past, events both real and imagined. Unlike Bachelard, however, these were never calming dreams for Walter; they were, from the very beginning, anxious, ambitious, inciting.

Building on Bachelard's conception of the home, pioneering cultural geographer Edward Relph explicates: "A place is not just the 'where' of something; it is the location plus everything that occupies that location seen as an integrated and meaningful phenomenon" (Relph 1973: 3). For Relph, this means that a place is always in process, never static, never stagnant; rather, "places are emerging or becoming" (ibid: 3). For Walter this means that the house is soon marked by failure and compromise, a bitter acceptance that his life never surpassed the level of 'a starter home'. To him, this also marks the family home as a trap, illustrating what Edward Relph refers to as "the drudgery of place," a paradoxical place which we are both committed to as "the very centres of our lives, but they may also be oppressive and imprisoning" (ibid: 41). Over the course of five seasons we watch Walter grapple with this paradox as he alternately longs to get away, fights to get back in (after his wife Skyler kicks him out), breaks into his own house on several occasions, before finally letting himself in one last time to the ruins of his home.

The first thing that Walter sees upon his arrival is that his house is now seemingly secured behind a padlocked fence marked with 'WARNING' signs by decree of the City of Albuquerque. However, Walter quickly discovers that the chain barely holds the gate together and he can easily slip past it – just as, no doubt, the skaters out back that opened the sequence also gained access. When Walter gets to the front door, we see that both the doorknob and locks have been removed, leaving only open holes in the door in their place. While the traditional living room is marked as a "deeply private place" and "the center of both family and individual life," as described by Relph in the mid-20th century (ibid: 36), in the 21st century the living room is recognized by contemporary theorists like Ensley F. Guffey as "a kind of liminal space and interface between the private and the social" (Guffey 2014: 159). Stripped of both door and lock – the objects that allow or deny access to the threshold, and allow the owners privatized access – the space of the living room is transformed to a public space for anyone who dares to crawl through the fence, i.e., anyone criminal. The fact that the word "HEISENBERG" aka Walter White's alter ego as criminal drug designer and czar is painted on the long, paneled wall that first welcomed us into the home when the young couple first visited the house, both signals that it is a space now marked for the criminal element and simultaneously marks the ruins of his family home as a significant historical location. Plotwise, of course this is no mere nostalgic return to the family home; the sentimental part of his journey is incidental. Walter's mission is to reclaim the vial of poison (ricin) he hid behind an electrical outlet in his infant daughter's room a season earlier which will prove vital to his wrapping up of loose ends by the closing of the series. In that sense, the ruined state of his house and home is a befitting welcome to his return with murderous intent. Drawing on theories of humanist geography developed in the mid-1970s, Ensley F. Guffey, in his essay "Buying the House: Place in *Breaking Bad*," argues that the "inevitable and increasing encroachment of Walt and Jesse's criminal behaviors into their personal and private lives" – which is such a major theme in the show – "is often shown by the intrusion of Jesse and Walt's illegal adventures into their respective homes" (ibid: 156).

Working within a framework of existential phenomenology first outlined by Maurice Merleau-Ponty in the mid-1940s, writers in the mid-1970s like Edward Relph and Yi-Fu Tuan argue that subjective human experiences give places meaning. While space can remain abstract, undefined and unclaimed, place is something specific, known, and often loved (or hated). Guffey differentiates between the use of "space" and "place" in *Breaking Bad*, arguing that our experiences within a space is what gives it meaning, importance, and significance which ultimately transforms it into place. As Guffey explains, "When we rent a new, unfurnished apartment, we often view it as an empty space, full of potential, perhaps, but still unrealized. As we move in, decorate, cook, spill drinks, argue, make love, get sick, become well, celebrate, mourn, eat, sleep, eliminate, and perform all the other ordinaries of human existence, the empty space becomes *our* space, our *place*" (ibid: 156). Over the course of *Breaking Bad* we watch as Walter's home is transformed to a place of crime; its hidden spaces like the crawl space under the house, the vents in his unborn daughter's room, even the unseen spaces behind electrical outlets are all repurposed to hide and house some aspect of his growing empire as the drug czar Heisenberg. In that sense the moral decay begins on the inside of the house long before it is visible on the outside, mirroring the cancer that is rapidly eating Walter himself away from the inside, invisibly and irreversibly.

The architecture of ruins: Golden lamps and crumbling homes

From an architectural perspective, the decay of buildings and their inevitable ruin provide unique and insightful reflections on our understanding of the human condition. For John Ruskin, writing in the 19th century, the ruin was a crucial eyewitness to demonstrate aging in architecture. Viewing architecture as a medium, Ruskin argued that a building must be conceived for both present and future use. As he described it, architecture had two duties, "the first, to render the architecture of the day, historical; and, the second, to preserve, as the most precious of inheritances, that of past ages" (Ruskin 1903 [1849]: 225). For that reason, he considered ruins noble, truthful, and tangible results of the passing of time.

For the spectator of *Breaking Bad* the ruins of Walter's house serve the same purpose(s). Glimpsed in flash-forward and outside of context midway through the final season in "Blood Money," they appear as a Dickensian warning of the inevitable future. When we witness them with Walter, seeing them for the first time, they are a brutal reflection of the past that is now gone. The family home has been erased. It is now marked as a scene both *of* crime and *for* crime (even if it is here just the diluted street crime of teenagers smoking pot and skateboarding in back). Writing in 1849, Ruskin's reflections on the impossibility of the sanctity of a home in ruins serve equally well here as a eulogy for Walter White:

"There is a sanctity in a good man's house which cannot be renewed in every tenement that rises on its ruins: and I believe that good men would generally feel this; and that having spent their lives happily and honourably, they would be grieved, at the close of them, to think that the place of their earthly abode, which had seen, and seemed almost to sympathize in, all their honour, their gladness, or their suffering, -- that this, with all the record it bare of them, and of all material things that they had loved and ruled over, and set the stamp of themselves upon – was to be swept away, as soon as there was room made for them in the grave; that no respect was to be shown to it, no affection felt for it, no good to be drawn from it by their children; that though there was a monument in the church, there was no warm monument in the hearth and house to them; that all that they ever treasured was despised, and the places that had sheltered and comforted them were dragged down to the dust." (ibid: 225-226)

Certainly this is an accurate descriptor of the shame and scorn heaped upon Walter and his family and that his family, in turn, also feel for their fallen patriarch – 'despised' and 'dragged down to the dust.' Functioning as mnemonic, the memories Walter confronts here includes the loss of his family, and subsequent loss of identities of father and husband. For Ruskin, it is the buildings that have withstood history, even though they be in ruins that are truly alive; marked by what he calls "that golden stain of time," a creation that is "more lasting than that of the natural objects around it"; these are "walls that have been witnesses of suffering, and its pillars rise out of the shadows of death" (ibid: 234). Here, the ruins rise up out of the shadows of the death of Walter White; the existing ruins reflect the truth about his life, bursting forth from the suburban disguise. The ruins of the house interrupt the seamlessness of the quiet neighborhood. As a space now marked both by and for crime it is a polar opposite to the 'Drug Free Zones' which are often constructed in a certain radius around public schools –

marked with signs indicating increased penalties for violations. Posing a threat to the surrounding neighborhood retroactively revealed, the house is set apart physically and symbolically by the fence put up around it, a space now reclaimed by the City of Albuquerque, as indicated by the WARNING signs posted on the fence and the front of the house.

Inside the house, however, the only sign we see is the spray-painted word "HEISENBERG" which Walter stops to consider upon entry. While the signs outside are warnings by the City of Albuquerque to keep law-abiding citizens out, the sign inside is a welcome to those who live outside the law and dare to push their way inside. It serves as a constructed resistance to the forces of ruin defining the continuing kingdom of Heisenberg – a regime of resistance to the law and the norms of society and sobriety itself – and a brand that will live on long after he is gone. If not a mark of immortality, it is certainly one of lasting ignominy. This functions not only within the diegetic universe of The ABQ as created by Vince Gilligan and his collaborators, but within the broader, international community of fans, where having access to the privileged knowledge that Walter White is Heisenberg (is Bryan Cranston) retains a certain currency of sub-cultural cool in spite of the fact that viewership is counted in the millions over a decade later. At the same time, the tag, to Walter White in that moment, also serves as something very much like a tombstone. For Walter White is no more; there is only Heisenberg.

The sociology of ruins, the materiality of decay

Pioneering sociologist Georg Simmel would build on the work of Ruskin in his 1903 essay, "The Ruin," recognizing the unique hybrid form of the ruin which fused the built environment with nature. As he describes: "The ruin of a building, however, means that where the work of art is dying, other forces and forms, those of nature, have grown; and that out of what art still lives in the ruin and what of nature already lives in it, there has emerged a new whole, a characteristic unity" (Simmel 1959 [1911]: 260). Simmel here is talking about the ruins of abandoned buildings, hence his description of nature once again regaining control over the space of the built environment. For Simmel this is not a case of "human beings destroy[ing] a work of man – but rather "that men *let it decay*" (ibid: 261).

In *Breaking Bad* we have been witness to the steady decay of Walter White's house and home for four and half seasons of television preceding this sudden flash-forward to the future. While the leap of undisclosed time can be roughly assumed to have taken us to the other side of his 52nd birthday (which we saw in flash-forward in the first episode of Season Five, "Live Free or Die" (07/15/12), due to his acquisition of hair on his head and face and the automatic rifle in the trunk), it isn't the giant Dickensian leap forward in time that the state of the house might suggest. This disjuncture, defying expectation, is of course a defining characteristic of the show – especially in its groundbreaking use of cold opens. At the same time, the commentary gleaned here is on the one hand arguing that the rot had already set in and was working on the house and home for a long time (indeed, for the duration of the show); on the other hand it is also a commentary on how quickly decay sets in in the world of crime. Viewed in this way, the quick ruination of the house is apt punishment for remaking the family home in this quiet suburban neighborhood into the hidden headquarters of Heisenberg. At the same time, it also holds up a mirror to Walter's quick and doomed battle with his own body as the cancer resumes its quick and deadly advance. There is also a third reading suggested by Simmel, who believed that every ruin is "an object infused with our nostalgia", which simultaneously marks the decay as "nature's revenge for the spirit's having violated it by making a form in its own image" (ibid: 259). Comprised of the Greek for both home (*nostos*) and pain (*algos*), nostalgia is a specific sort of pain one feels for what was known but now lost. Addressing Simmel's claims in a 21st century context, Andreas Huyssen refers to this longing for a lost past as "the modern disease per se," pointing out the architectural ruin as a common trigger, explaining: "In the body of the ruin the past is both present in its residues and yet no longer accessible, making the ruin an especially powerful trigger for nostalgia" (Huyssen 2006: 7). For Walter, as for the viewer, seeing the family home in such a state triggers memories of earlier, happier days – and for the viewer, possibly even waves of nostalgia for those earlier episodes when it looked like Walter might get away with it (if/when we still wanted that to happen).

Writing in the 21st century, sociologist and cultural geographer Tim Edensor admits in the introduction to *Industrial Ruins: Spaces, Aesthetics and Materiality* that his book "evolved out of his enthusiasm for visiting industrial ruins," opening the book with a picaresque recalling of his childhood romance with local ruins, constructing them as a sort of primal scene which has marked him and provides the inspiration for his research. (Edensor 2005: 7.) This also grounds the book in issues of nostalgia, which connects with Simmel's thesis, but has also earned him criticism by peers within sociology and outright scorn by a number of architects. As Edensor describes in his introduction, "one of the main objectives of this book is to contest the notion that ruins are spaces of waste, that contain nothing, or nothing of value, and that they are saturated with negativity as spaces of danger, delinquency, ugliness and disorder" (ibid: 7). At first glance this seems to align him with Ruskin, but to Edensor (channeling Henri Lefebvre), the value of these ruins lies in the fact that they can serve as "the site of current or future production," which can be "exploited for profit" (ibid: 8). Part of the impetus for transforming ruins into usable space stems from the common association of derelict space with deviant behavior, which in turn promote fears of violence and crime.

Edensor's insights regarding the relationship of ruins and property development are informative when considering the ruins of Walter's house – which erodes at a lightning-fast pace, even eliding the conceit of time in the televisual universe. As Edensor describes: "if spaces are conceived as disturbingly non-functional, they must be replaced and filled in – turned into abstract space – to remove these signs of unproductive and unfunctional blankness. Frequently, they are asset-stripped and then cleared to encourage property speculation because dereliction appears as a scar on the landscape composed of matter out of place, which must be erased and then filled in with something more 'useful'. (ibid: 8). In that sense the quick transformation of Walter's house to ruins is a necessary part of the process of redefining the home; like pressing restart. As Edensor sees it,

"The designs and schemes of modernist planners to rationalise the landscape are embodied in the structure and organisation of industrial and domestic techniques for spatial control which form part of the 'machinic episteme', the belief that an all-encompassing design can order meaning through the logical placing of people and things within a grid-like system. The grid delineates the functions of specific areas so as to produce a series of single-purpose spaces where preferred activities occur, creating what Berman terms a spatially and socially segmented world – people here, traffic there; work here, homes there; rich here; poor there." (ibid: 54)

In *Breaking Bad*, the reveal of Walter White's secret identity as drug czar Heisenberg also reveals that the family home has been operating outside of its delineated use-function: a home is no longer a home, but an important hub of the illicit drug trade. All previously assumed categories are subverted: drug traffic instead of the staticity and solemnity of the home; work instead of leisure; the (secretly) rich (at least for a while) instead of poor. Considered in this way, the transformation of these categories would be as fast as word of mouth, also explaining the rapidity of its transformation to ruins. This space suddenly marked as 'impure,' violating what the city and local neighbors mark as violating the existing social contract, is to be closed off from the surrounding area, now surveilled, monitored and controlled by the carceral apparatus. Certainly part of the way the City of Albuquerque marks Walter's former house as impure or criminal is by cutting off its access to the privileges of a city's infrastructure: lights, power, water. This, in conjunction with the stripping of doorknobs and locks helps to further break down the differentiation between inside and outside as well as public and private; in a sense, returning the space to nature. This rite of purification serves as both punishment for Walter's crimes and preparation for whomever lives there next. As essentially a hard restart, the process will require both a remodel, and the subsequent re-entry of the house into the housing market, itself a long-held form of ritual that will transform the public back to private, and the ruins of Walter's house back to a (new) family home. It is only after going through this process that the house can re-emerge and resume its place in the grid of the neighborhood, returning the space to its original, intended use-value as a family home.

Ruin porn, tourism, conclusion

Enlarging the scope from the ruin of a building, or even neighborhood, to the scale of ruin-cities we immediately think of Chernobyl, a post-war Hiroshima, or even Detroit. But Vince Gilligan's vision of the city of Albuquerque as "The ABQ" resonates on similar levels. The intimate and heart-wrenching access we are given to what was once Walter's family home is an insider's glimpse of the rot that is spreading – through the city, through America, and through Western society. In that sense, the ruins of Walter's house are part of a culture of ruins we have encountered and inhabited throughout the series. Vince Gilligan's vision of the built environment of Albuquerque is a city stippled with ruins. We see Jesse's house fall victim to the same fate. Twice. When Jesse's girlfriend Jane dies in Season 2 in the episode tellingly titled "ABQ," Jess relocates to "The Shooting Gallery," the street name for a run-down trap-house populated by heroin addicts, scoring, shooting and sleeping. One of the other most conspicuous form of ruins (in the sense of humanity and the built environment is of course the Crossroads Hotel (aka "The Crystal Palace," as Walter's brother-in-law and DEA agent Hank Schrader christens it, tongue in cheek, in Season 1) – a constant grounding location on the show, same as it is for commuters passing by it daily on the freeway now. Or the corner where Combo was shot. Or Tuco's headquarters. Or the junkyard where "The Crystal Ship" – i.e. Walter and Jesse's RV (motorhome/meth lab) – was parked. Or the railyard. The list goes on. (And on.)

For this reason, I would describe Gilligan's portrayal of Albuquerque as a tele-visual example of "ruin porn" – a phrase credited to Detroit blogger and photographer, James Griffioen, now a common term to describe the photography of post-industrial ruins, now a staple on the internet, and subject of a large number of coffee table books in the past decade. The term first appeared in a 2009 *Vice* magazine article by Thomas Morton, who was in Detroit writing a piece on the popularity of the city's urban decay. (Morton 2009) Ruin porn is often derided as an exploitative vison of a community's suffering. As the 'porn' in 'ruin porn' suggests, the enjoyment it inspires is scopophilic access to that which shouldn't be seen. Indeed, much of the ongoing appeal of the show is seeing that which shouldn't be seen – often taking the form of ruins – whether it is the ruins of a house, home, kingdom, or person ☐ the ruins behind the façade; the ruins that wreck a neighborhood; the ruins that lie in wait. Simultaneously a fantasy of destruction and a mournful nostalgia that Kate Brown terms "rustalgia," our enjoyment of ruin porn oscillates within the poles of this circuit. (Brown 2014). Certainly this is at work at various times throughout both *Breaking Bad* and *Better Call Saul* in their portrayals of Albuquerque as post-noir and, at times, seemingly post-apocalyptic.

The ruins we encounter in both *Breaking Bad* and *Better Call Saul* speak to the past, present and future of Albuquerque – illustrating what Siobhan Lyons describes are the defining characteristics of ruin porn: "They remind us, in a very sublime way, of the inevitability of human extinction, refocusing the terrain of 'ruin' away from the ancient world and towards the imminent future" (Lyons 2018: 1-2.). For Lyons "ruin porn is so compelling precisely because it is a bewildering form of time travel to the future within the present" (ibid: 1-2.). Lyons goes on to argue that "ruin porn acts as a kind of 'dark tourism,' its affiliation with death proving irresistible to a world tiring of the increasingly 'hyperreal' conditions of everyday tourism" (ibid: 6). Termed elsewhere as "traumascapes" by Maria Tumarkin (2015) or "obscenery" by Iain Sinclair (2002), ruin porn both exploits and romanticizes decay while at the same time feigning intimacy and offering a form of jouissance that exists outside what many consider accepted norms. It also relegates enjoyment at a certain distance, accessible primarily in the scopic regimes. Similarly, Tim Edensor claims that "exploring a ruin is a kind of antitourism," arguing that "There are no obvious spectacles around which to organise a tour, or which fit into expectations about what will be gazed upon, and sights will often be indecipherable. There is nothing to buy and nothing conforms to the staged aesthetics of tourist space. These experiences cannot be inserted into a pre-arranged vocabulary or classified as 'exotic' or 'typical'" (Edensor 2005, 95)

But in in our lived-world in the city of Albuquerque, they are. The list of ruins named above are all stops on the "Breaking Bad RV Tour" – which tourists can experience as part of a three-hour journey past more than twenty locations from the show(s) in an RV identical to the "Crystal Ship" in the show. Here, ruins, residences, restaurants and other key locations from the show are experienced with equal glee from tourists arriving from all over the world. Tellingly, the program is credited with increasing tourism in Albuquerque over the past decade – especially the demographic of 20-50-year olds – which, no doubt, is also due to the subsequent popularity of the prequel, *Better Call Saul*, which aired its sixth and final season on AMC in 2022 – and to the continued high-profile presence of both shows on Netflix.

In our lived-world the real life house of fictional character Walter White remains a top tourist attraction in Albuquerque, still attracting a heavy flow of traffic by the house, and a steady stream of pizzas on the roof of their garage. It was these pizzas that attracted national media attention in 2015 (on CNN and Time Magazine) and prompted a plea to fans by the show's creator, Vince Gilligan, to stop – which was ignored. Finally, in 2017, the Quintana family who has lived there for 40 years (kindly allowing the crew to shoot there) erected a six-foot-tall steel fence. It was a difficult decision for the family, Joanne Quintana told local Albuquerque news station KOB4 reporter: "We don't want to gate ourselves in. We're the ones who are being locked up." (Bruner 2017)

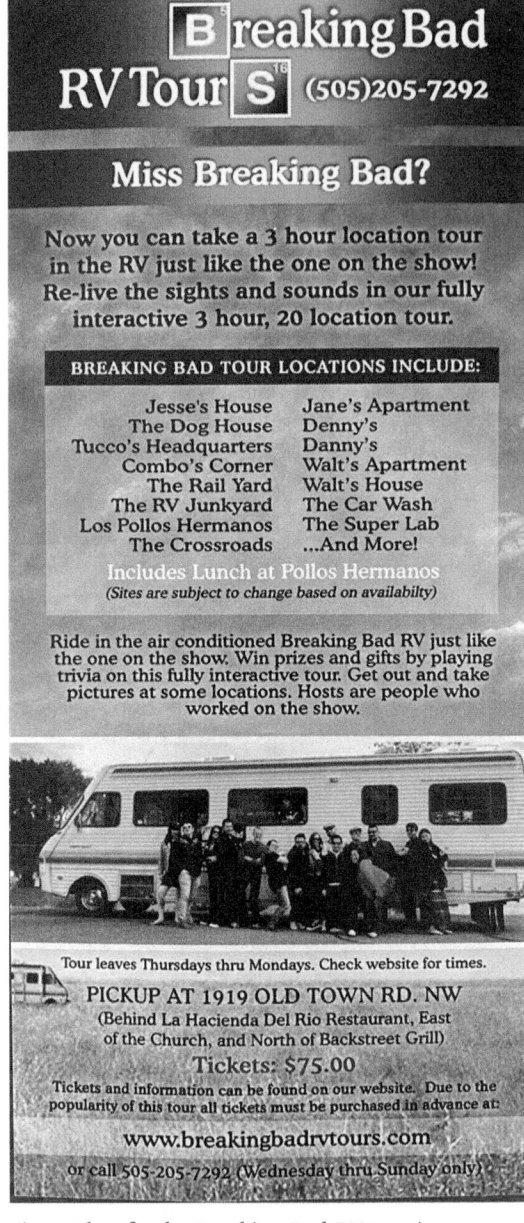

Fig. 1: Flyer for the Breaking Bad RV tour in Albuquerque, New Mexico (front).

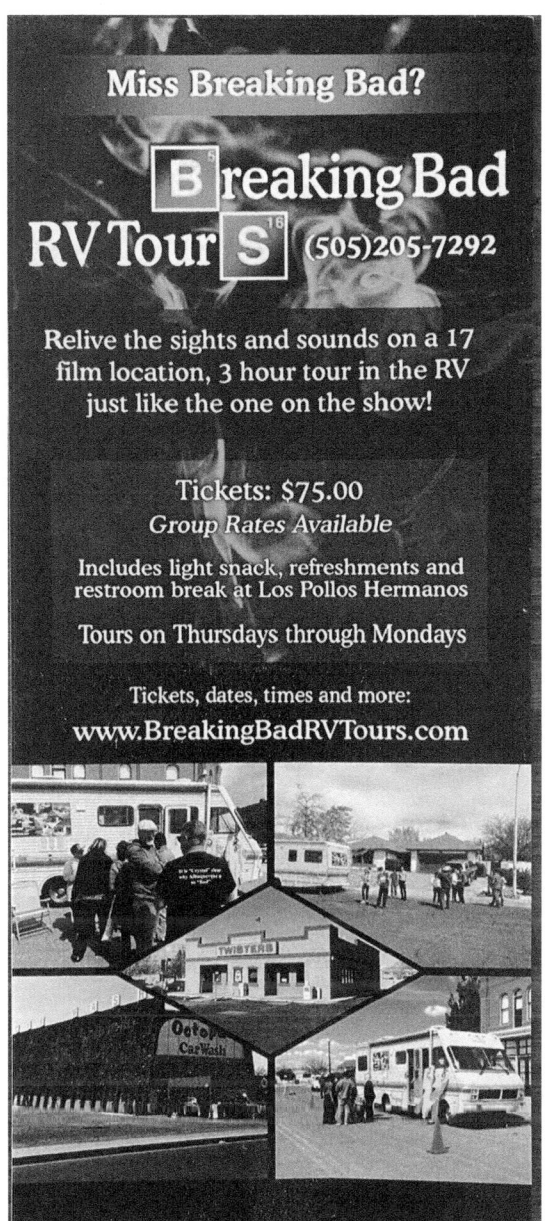

Fig. 2: Flyer for the Breaking Bad RV tour in Albuquerque, New Mexico (back).

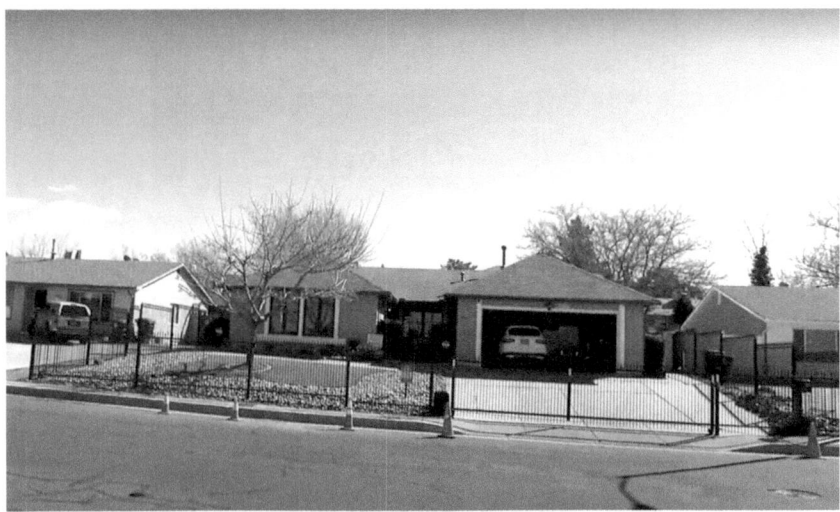

Fig. 3: Outside the real life house of fictional character Walter White in 2015.

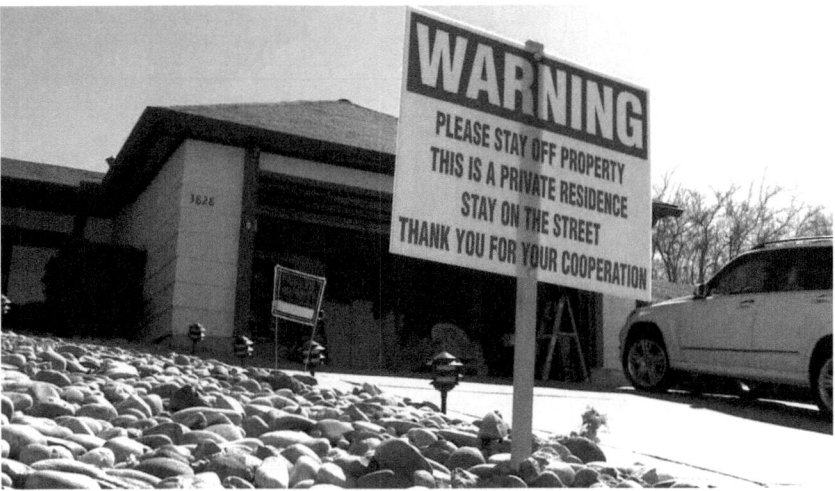

Fig. 4: Outside the real life house of fictional character Walter White in 2022.

With the addition of the fence, the house is again marked as ruin. Ruined by the very tourist industry that the popularity of the show prompted; ruined, at least in part, by the ruin porn that the show presents. With the intention of keeping fans out, the inhabitants of the house are also, effectively, sealed in *and* separated from the rest of the neighborhood. At the same time, the fence acts as a frame around the property, calling attention to the strange space which is no longer their own, no longer private, no longer suburban, no longer existing entirely either in the space and time of the show, or of the here and now.

References

Bachelard, G. (1994 [1964]): *The Poetics of Space*. Maria Jones (trans). Boston, MA: Beacon Press.

Brown, K, (2014): *Dispatches from Dystopia: Histories of Places Not Yet Forgotten*. Chicago: University of Chicago Press

Bruner, R. (October 11, 2017): "*Breaking Bad* House Owners Put up Fence Because People Can't Stop Throwing Pizzas," *Time*. Retrieved from https://time.com/4977859/breaking-bad-house-fence/

Edensor, T. (2005): *Industrial Ruins: Spaces, Aesthetics and Materiality*. New York: Berg.

Genette, G. (1983): *Narrative Discourse: An Essay in Method*. Jane E. Lewin (trans). New York: Cornell University Press.

Guffey, E.F. (2014): "Buying the House: Place in *Breaking Bad*." In David P. Pierson (ed.) *Breaking Bad: Critical Essays on the Contexts, Politics, Style, and Reception of the Television Series*. New York: Lexington Books, pp. 155-172.

Huyssen, A. (Spring 2006): "Nostalgia for Ruins." *Grey Room* 23, pp, 6-21.

Lyons, S. (2018): "Introduction: Ruin Porn, Capitalism, and the Anthropocene." In Siobhan Lyons (ed.) *Ruin Porn and the Obsession with Decay*. London: Palgrave Macmillan, pp. 1-12.

Morton, T. (July 31, 2009). "Something, Something, Something Detroit," *Vice*. Retrieved from https://www.vice.com/en/article/ppzb9z/something-something-something-detroit-994-v16n8.

Relph, E. (1973): *Place and Placelessness*. London: Rion.

Restivo, A. (2019): *Breaking Bad and Cinematic Television*. Durham, NC: Duke University Press.

Ruskin, J. (1903 [1849]): *The Seven Lamps of Architecture* in *The Complete Works of John Ruskin, Library Edition, Volume VIII*, E.T. Cook and Alexander Wedderburn (eds), London, George Allen.

Simmel, G. (1959 [1911]): "The Ruin." In: David Kettler (ed.), David Kettler (trans), *Georg Simmel, 1858-1918: A Collection of Essays, with Translations and a Bibliography*, Columbus, OH: Ohio State University Press, pp. 259-266.

Sinclair, I./Habersham, E.M. (2002): *White Goods*. Uppingham: Goldmarks.

Tumarkin, M. (2015): *Traumascapes*. Melbourne: Melbourne University Press.

"A Beachfront Property in Gaza is Not a Dream"

Utopia, the Digital, and the Image of the Ruin in Palestine

Daniel Vella

Abstract

This essay takes as its starting point an image shared in December 2023 by the Israeli real estate firm Harey Zahav, depicting the plans for a row of luxury beach villas digitally superimposed onto a photograph of ruins in Gaza. Taken in the context of other images of ruination which have been shared and remediated during the assault on Gaza following October 7, 2023, this essay argues that the image is emblematic of a radical shift away from the long-standing Israeli practice of clearing or otherwise obscuring Palestinian ruins within the territory of Israel, as a means of concealing both the pre-1948 Palestinan presence in the land and the violence and ethnic cleansing through which the state was founded.

Drawing on discussions of utopia by Ernst Bloch, Fredric Jameson, China Miéville and others, the significance of the representation of images of Palestinian ruins shall be read this in the context of the idea of utopia, and the 'forced forgetting' of the foundational violence upon which the concept of the utopian 'blank slate' is founded. The essay shall then build on the work of Saree Makdisi, Ilan Pappe, Walid Khalidi and others to survey the history of the Israeli treatment of Palestinian ruins, and the way in which their active destruction has been a key mechanism in a process of memoricide at work both in the Israeli relation to the landscape and discursive practices. Finally, the essay shall consider how recent images like the one produced and shared by Harey Zahav represent a shift towards a foregrounding – rather than a concealment – of state violence in the Israeli-Palestinian context, and read the significance of this shift in relation to what Taner Akçam, following Hannah Arendt, has referred to as the "end of tradition" (2025) of the post-war Western liberal world order.

Keywords

ruins, Gaza, Israel, utopia, Zionism

In December 2023, two months into the assault on Gaza that followed the Hamas-led atrocities of October 7, the Israeli real estate firm Harey Zahav – which specializes in building property in the occupied West Bank – posted an image on their social media channels which showed the plans for a row of luxury beach villas, digitally rendered and superimposed onto a photograph of bombed-out ruins along the Gaza coast. The tag line for the image read: "Wake up, a beach-front property is not a dream."

The image was widely shared, leading to an international outcry that Harey Zahav addressed by claiming the social media posts were intended as a joke (Dayan 2023). Whatever the intention behind the image, though, its creation and dissemination by a corporate entity seems to signal a radical shift in how the recent – one could even say present-tense – ruins of conflict are imaged, mediated and granted significance.

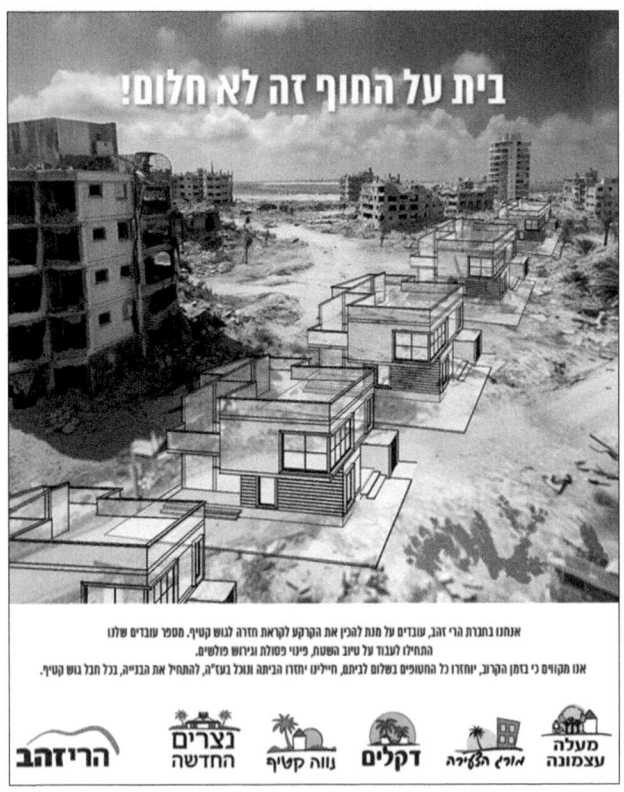

Fig. 1: The image shared by Harey Zahav, showing the plans for luxury beach villas superimposed upon the ruins of Gaza. (Image: 2023 Harey Zahav)

The war-ravaged city in ruins is a familiar trope in both documentary and fictional representations of conflict. In recent years, the documentaries *For Sama* (Al

Kateab/Watts 2019) and *20 Days in Mariupol* (Chernov 2023), for example, have foregrounded images of, respectively, the city of Aleppo in ruins in the midst of the Syrian Civil War, and the city of Mariupol in the first weeks of the Russian invasion of Ukraine. However, these are images produced by those who live in the ruins, or those who sympathize with them. Their rhetorical intent, broadly speaking, is to bear witness – to draw attention and invite condemnation of the violence that made the ruins, or to instil compassion towards those who have been, or continue to be, the victims of that violence.

Such ruin-imagery also exists in the context of the assault on Gaza, of course. To take only one of the more recent examples at the time of writing, on July 22, 2025, the UK pressure group *Led by Donkeys* draped an image of the ruins of the Jabaliyah refugee camp on the building opposite the headquarters of the British Labour Party in London, in a criticism of the UK government's criminalizing of pro-Palestinian protest.

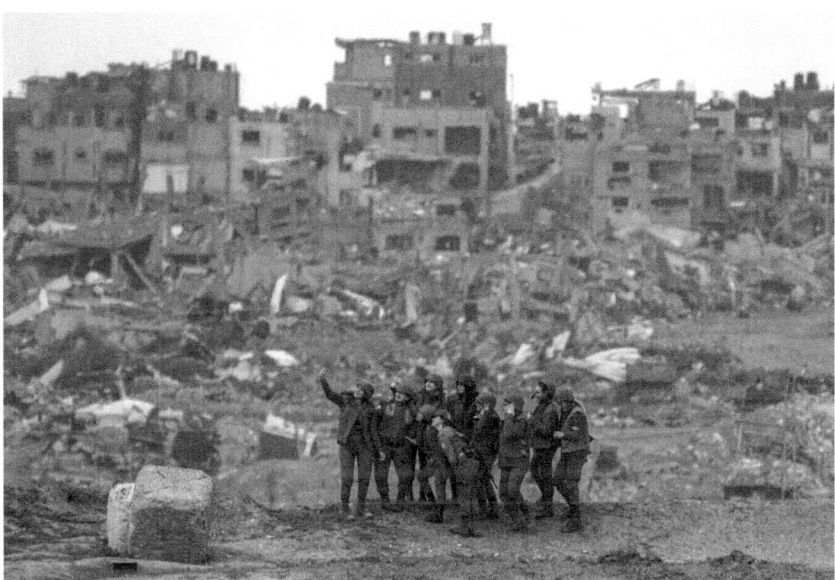

Fig. 2. A photograph by AP photographer Tsafrir Abayov, showing IDF soldiers taking a selfie against the ruins of Gaza. (Copyright: 2024 Associated Press)

The image shared by Harey Zahav, though, operates in a different rhetorical register. The intent of the image is not to invoke empathy or pity, or even anger. It aligns, in a way, with another much-circulated photograph from the same month – the one taken by Associated Press photographer Tsafrir Abayov of a group of IDF soldiers taking a selfie as they stand in front of the ruins of Gaza (Abayov 2024). Rather, the connection we can make is not to Abayov's photograph, but to the image we can presume the soldiers are taking – the selfie in which they position themselves, with celebratory pride, against the ruination they have had a part in causing.

In this essay, I shall argue that such ruin-images represent a radical shift away from the long-standing Israeli practice of clearing or otherwise obscuring Palestinian ruins within the territory of Israel, as a means of concealing both the pre-1948 Palestinan presence in the land and the violence and ethnic cleansing through which the state was founded. I shall read this in the context of the idea of utopia, and the "forced forgetting" (Miéville 2016) of the foundational violence upon which the concept of the utopian 'blank slate' is founded. In relation to this, I shall survey the history of the Israeli treatment of Palestinian ruins, and the way in which their active destruction has been a key mechanism in a process of "memoricide" (Pappe 2006) at work both in the Israeli relation to the landscape and discursive practices. Finally, I shall consider how recent images like the one produced and shared by Harey Zahav represent a shift towards a foregrounding – rather than a concealment – of state violence in the Israeli-Palestinian context, and read the significance of this shift in relation to what Taner Akçam, following Hannah Arendt, has referred to as the "end of tradition" of the post-war Western liberal world order (2025).

Utopia and ruins

Intuitively, ruins and utopia seem like antinomies. Ruins cast our glance backwards; utopia makes us look foward. Utopia announces itself with an apparent singularity of purpose, meaning and vision, while ruins are, by their very nature, polyvalent – as Salvatore Settis maps out in this delineation of all the different lines along which a ruin can be read:

In their persistent presence, ruins speak to us of the structures they once were, of the people who made them, of those who commanded them to be made, and of those who for a time made use of them. In their evocation of absence, they speak of those who destroyed them or abandoned them or failed to protect them from the irresistible ravages of Time. In their present state, ruins speak of those who have tried to make sense of them, or have been drawn to represent them, or have used them as objects of memorialization. (1997: vii)

Yet, ruins are everywhere we look when we look at Utopia: ruins populate the shadows cast by utopian light. And this is despite the fact that utopia has often been cast precisely as the escape from ruination and the burden of historical memory. We need only pause upon a brief consideration of that utopian promise which is so foundational to the West that it arguably gave the West its name: the promise of the so-called New World. Looking across the Atlantic, Goethe once wrote:

America, you have it better than our continent – the old one.
You have no tumbledown castles
And no basalt deposits.
Your inner lives are not disturbed by
Useless memories and vain strife. (1813)

Goethe's implicit contrast is with a Europe that is old, and scarred by the marks of conflict both political – the tumbledown castles – and geological – the basalt deposits: the Europe that is the stomping-ground of Walter Benjamin's Angel of History, whose gaze, fixed backwards, sees only "one single catastrophe which keeps piling wreckage upon wreckage and hurls it at its feet" (2003: 392). Goethe positions America as the utopian antithesis to all this – a blank slate, a fresh start. Of course it was no such thing even then, let alone now. The expansionist, settler-colonial ideology of Manifest Destiny required the dehumanisation and the erasure of the indigenous populations and cultures who already called the lands of the so-called New World home (Madley 2015). In many parts of North America – especially in the southwestern United States – the ruins of Native American cultures haunt the landscape, disproving the narrative of a vast wilderness at best sparsely populated by wandering primitive tribes (Alvarez 2014: 2). Moreover, Goethe's fresh new continent is now littered with its own ruins. Martin Procházka writes that "American ruins often function as subversive heterotopias, unsettling discourses of progress and other ideological views of history" (Procházka 2012: 72). From the deserted ghost towns of Western expansionism to the post-industrial decline of Detroit, ruins in what we still call the new world represent the failure of a number of American utopias: the utopia of the Western frontier, the industrial utopia of a booming post-war manufacturing economy – and, all in all, the Manifest Destiny of America as the culmination of history, whether this was understood to have a theological or humanist basis.

Still, the kind of utopian thinking that dreams of a clean slate and a fresh start – that imagines a lifeworld free of the rubble of history – endures, even if it keeps having to find new geographical playgrounds. Daniel Herwitz writes of the twentieth-century "impulse to fill the so-called empty spaces of the third world with modernist masterpieces and utopian, modernist cities" like Brasília in Brazil or Chandigarh in India (2010: 235). In these places, where planned cities intended to serve as national or regional capitals were constructed according to top-down designs, "power seems to find a clean slate, but only because it fails or refuses to notice what it has cleared away in its grand modernist sweep" (ibid.).

There is always, in the imposition of such large-scale, top-down, planned architectural projects, an uneasy coexistence of what power speaks about – the new social order which is being promised and set in stone – and what it does not – the existing socio-cultural context which has to be swept aside for it. The socialist planned city of Nowa Huta laid out on the outskirts of Kraków in Poland in the years following the Second World War "would represent a new kind of civilization, peopled by 'new men' (Lebow 2013: 14). The intent was specifically to lay out a new city that embodied the Communist values of the Polish People's Republic, in a challenge to the largely middle-class, bourgeois character of Kraków's population. However, the site on which Nowa Huta was laid out was no blank slate. A number of villages had to make way for the new construction. Moreover, as the new city was being built, "archaeologists found [...] evidence of human settlement dating back to Paleolithic times," as well as – in an almost too apt mythic echo of national resistance to foreign influence – "'Wanda's Mound,' a 14.5 metre high, most likely pagan, burial mound, said to contain the remains of Queen Wanda – she who had thrown herself from the ramparts of Wawel Castle to avoid marrying a German prince" (ibid: 26-27).

This tension between the utopian promise of the radically new – together with the idea of the clean slate on which it is predicated – and the seeming necessity of the violence inherent in its creation is one which lies at the heart of the idea of utopia itself. In Thomas More's 1516 *Utopia* – which gave all future utopias their name – it is striking to note that the Utopians were not the indigenous inhabitants of the mythical island. Instead – though this detail is all but glossed over – the island upon which the Utopians build their perfect civilization is theirs by conquest. China Miéville notes that:

We don't know much about the society that Utopus and his armies destroyed – that's the nature of such forced forgetting – but we know its name.... "for Abraxa was its first name." We know the history of such encounters, too; that every brutalised, genocided and enslaved people in history have, like the Abraxans, been "rude and uncivilised" in the tracts of their invaders" (2016: 6-7)

Here, the erasures and dehumanisations upon which the Utopian clean slate depends align with Frantz Fanon's description of the logic of settler-colonialism, according to which "the settler paints the native as the quintessence of evil" (1961: 32), dehumanising the native population to the point that settler-colonial violence is painted as not only practically but also morally necessary. The native "represents not only the absence of values, but the negation of values" (ibid.) – indeed, the native becomes an avatar of everything civilized society is not: savage where it is civilized, lawless where it is lawful, ignorant where it is enlightened, primitive where it is progressive, irrational where it is rational, lazy where it is hard-working, and so on. There is, then, a sense in which Utopia needs the idea of Abraxa, at the same time as it requires Abraxa to be swept aside. It is enough to recall Edward Said's delineation of the structure of Orientalist thought, by which the West defines itself negatively in distinction to its own construction of the East: "the Oriental is irrational, depraved (fallen), childlike, 'different'; thus the European is rational, virtuous, mature, 'normal.'" (2003: 40).

Utopia, then, might well be a structure of hope – an "anticipatory consciousness," as Ernst Bloch put it, gesturing towards alternative political, social and cultural arrangements that have not yet taken shape, but that could one day do so (1986). At the same time, though, utopia points as much towards – and is defined as much by – what has to be swept aside as what it aims to build. As Fredric Jameson wrote, "the Utopian remedy must at first be a fundamentally negative one [...] [utopias] seem to offer blueprints: these are however maps and plans to be read negatively, as what is to be accomplished after the demolitions and the removals" (2005: 12). For utopia to retain its utopian character, though, this violence needs to be rendered invisible. If the utopian society is to represent the ideals it strives for, then it must perform, to return to Miéville's term, a "forced forgetting" of its genesis.

In this context, then – to borrow, briefly, a psychoanalytic term – ruins are the uncanny in the utopian project: the return of the repressed and the return of the same (Royle 2003). They signify the endurance of the memory, not only of that which utopia sought to differentiate itself from and defined itself against, but also of the violence of its foundational gesture. The ruin complicates utopian univocality, giving the lie to its no-placeness and returning history into the picture.

Israel, Palestine, utopia and ruins

In his survey of utopian vectors in *fin-de-siècle* Mitteleuropean Jewish thought, Michael Löwy marked an "elective affinity" between "Jewish messianism" and "libertarian utopia" (2017: 26) as being constitutive of the Judeo-Germanic intellectual culture that produced thinkers as diverse as Franz Rosenzweig, Franz Kafka, Ernst Bloch, Martin Buber and Walter Benjamin. There was no unified stance on Zionism as an intellectual or political project within this milieu – many of these thinkers, in fact, either explicilty rejected the Zionist project or were ambivalent about it (ibid.). Nonetheless, it was agains the background of this intellectual context that Zionism emerged, and its vision was, in at least some of its versions, distinctly utopian in character – an "anticipatory consciousness," to return to Bloch's term.

Theodor Herzl, founder of the Zionist Organization, expressed this vision in the 1902 novel *Altneuland* (translated into English as *Old-New Land*). Closely adhering to the utopian generic structure and form (Stolow 1997), the narrative has its protagonist, the Jewish Viennese lawyer Friedrich Löwenberg, visit Palestine twice – once in the present of 1902, and again, following twenty years of isolation on a South Pacific island, in the future of 1923, where he encounters a land entirely refashioned by Jewish settlement and stewardship into a technologically advanced, socially progressive society of modern, rationally laid out cities and thriving industry. There are, of course, major divergences between the utopian Zionist society Herzl imagines in *Altneuland* and the state of Israel which would become a reality decades after his death (Robinson 2013), and yet the idea of Israel has often continued to be presented in similar utopian terms.

Zionist discourse has often spoken of Palestine as "a land without a people for a people without a land" – the slogan was used, for instance, on the Israeli pavilion at the 1958 World's Fair in Brussels (Azaryahu 2004). However, if the idea of a blank slate and an empty territory waiting to be claimed was already untenable in relation to the vast expanses of the American West or Australia, with their comparatively sparse indigenous populations, in Palestine "the process occurred in a tiny and already relatively densely settled country where there could have been no perception of a vast untapped wilderness crying for Western exploration and exploitation" (Khalidi 1991: xxxi). The landscape, in fact, bore the inconvenient visible marks of inhabitation by a people of whose presence it needed to be cleansed.

Amir Eshel observes a duality in the Israeli relation to ruin. The founding of the city of Tel Aviv in 1909, he writes, was "motivated by the desire to build a clean, European neighbourhood on the outskirts of Jaffa, which was seen as grimy and Oriental" (2010: 137). In the gesture of its foundation, Tel Aviv had the character of the aforementioned twentieth-century Utopian planned cities that would succeed it: Brasília, Chandigarh, Nowa Huta. It even anticipated the modernist architectural principles of the former two: the White City, a collection of over 4,000 buildings constructed in the 1930s, remains the largest grouping of buildings in the Bauhaus/International Style anywhere in the world.

Still, this utopian promise of the new was only one side of the discourse that accompanied the founding of this new city. Eshel writes that "in Hebrew, *tel* is a mound of ancient ruins, while *aviv* means the season of spring. The name of the new city represents the claim that it was built on the ruins of ancient Jewish civilization in the promised land" (ibid.). In fact, 'Tel Aviv' was also the name given to Altneuland in its Hebrew translation by Nahum Sokolow, further aligning the city with Zionism as a utopian project. The etymology of the city's name, then, shows a double temporal face: a future utopia founded upon the renewal of past glory, in an implicit connection of ancestral, almost mythical past and promised futurity that erases the centuries of history separating one from the other.

This is a familiar relation to ruin: the idea of ruin as heritage, a term both conceptually and etymologically laden with the suggestion of birth right and inheritance. It is similar, for example, to the classicism that has defined the Italian relation to the ruins of Rome from the Renaissance (Riegl 1982) through to the nineteenth-century nation-building project of a unified Italy (Kirk 2008) and Mussolini's fascist vision of a new imperial Italy (Notaro 2000). In all these times, the ruin has served as the material correlate of the purportedly essential – but ultimately ideological – connection and continuity between a grand, heroic past and the nation-building project of the given time.

However, the way that ruins "serve in contemporary Israel as a central trope in fostering a sense of community through a connection to the ancient Jewish past" (Eshel 2010: 137) is shadowed by another, more troubled relation to ruin. Eshel writes about growing up in Haifa, a city marked by a certain landscape of ruins that was not a part of this heritage project of ethnic and national community-building. In the city of his youth, he says, there were ruins that were visible but not acknowledged. "The deserted, cemented-up houses in the Arab quarter were left silent," he writes, "no sign explained their haunting stillness" (ibid.). In fact, the endurance of the ruins seems to give the lie to "the usual story told about the emergence of present-day Haifa," which, in typically utopian fashion, is "one of creation out of nothing" (ibid.).

The enduring presence of the abandoned houses of the city's former Palestinian residents, then, complicated the utopian narrative of the city and the nation. Nor are these the only such ruins in the Israeli landscape. Saree Makdisi has discussed at length the vexed relation of Israel to the landscape of Palestinian ruins left behind by the Nakba of 1948 – and by the other waves of ethnic cleansing since then. After the forced displacement of the Palestinian population, Makdisi writes, "the new Israeli state controlled a desolated terrain: a Palestinian Arab landscape of (now) empty and partially ruined villages" (2022: 20). These ruins presented a problem: they stood as reminders of the Palestinian presence in the land that preceded the state of Israel, as well as of the violence of its displacement.

To address this, as Aron Shai writes, in 1965, "a clear policy was established to "level" the abandoned villages with the aim of "clearing" the country, to quote the official term used at the time" (2006: 87). This was seen as a necessity, among other reasons, because the ruins were considered problematic for the image Israel wanted to present of itself: it was vital to "stop tourists from raising, as one Foreign Ministry official put it [at the time], "superfluous questions" about the ghostly landscape of ruins inhabited by Israelis" (Makdisi 2022: 21).

Cleansing the landscape from the ruins of Palestinian life, then, fulfilled an important function for the then-nascent Israeli state: it removed, from the eyes of the international community, the evidence that would have countered the narrative of a 'land without a people.' Just as important, Makdisi argues, "the idea was that the erasure of the landscape would erase with it the political rights and claims based on belonging" (ibid: 21-22) of the displaced Palestinian population. As Rochelle A. Davis writes, "the state's policy during and after the 1948 War was to destroy the houses in the villages so that people would not be encouraged to return." (2011: 9).

This erasure occurred on symbolic as well as material levels. Comparing Israeli and Palestinian maps of the region over the past decades, Jess Bier notes that significant efforts were made to obscure the presence of Palestinian communities on Israeli maps – a process which, conversely, left its own negative traces in the symbolic record: "Israeli population cartography incorporated Palestinians through the very act of erasing them" (2017: 83). On the 1992 graduated circle population maps of the Israeli geographer Roberto Bachi, for example – which counted Israeli but not Palestinian residents – "the erased Palestinian populations are so numerous that they show up precisely through their absence, as a white, blank area" (ibid: 103). Jameson's point about utopias being maps that are to be read negatively could scarcely have a more literal illustration.

Ahmed Moor refers to the "manufactured nonexistence of the Palestinians" (2024: 17). Thus is created the blank slate, the empty territory upon which Utopia can be built. (As an aside, it is worth pointing out that – as Muhammad Ali Khalidi observes – though in Herzl's *Altneuland* the Zionist utopia is a multicultural one, with Palestinian Arabs continuing to live within the Jewish-led future society alongside people of all other cultures, the Palestinian villages which so offended Löwenberg's European sensibility on his first voyage to Palestine are gone – "the Arab villages have been removed from the landscape" (Khalidi 2001: 58).) Here we have, to return again to Miéville's term, a "forced forgetting" – the Palestinians are the Abraxans of the Zionist Utopia. There is a "colonial logic" at work, Makdisi says: the ruins serve as a metonym for an entire natural and cultural landscape that is Othered and deemed worthless, barren and undeveloped. In Zionist discourses, the native Levantine landscape is depicted as a wasteland to be cleared to make way for the industrialised agriculture and pine forests of the settler-colonial imagination: "settlement brings order, system, and modernity to a backward, chaotic, unproductive, unsettled indigenous landscape of barren 'neglect'" (2022: 20).

In fact, one of the primary mechanisms for the erasure and the forced forgetting of the ruins and the Palestinian landscape they made part of was the extensive afforestation project undertaken by the Jewish National Fund – which, Makdisi notes, has replaced this native cultural landscape and ecology with a landscape more pleasing to the European settler-colonial eye, but ecologically unsuited to the climate or the native biodiversity (ibid: 32).

Makdisi's point is not just that this erasure happens, but that there is an intrinsic dual structure of affirmation and denial at work. The cleansing of the land from all traces of pre-1948 Palestinian life was not presented as what it was – an act of ethnic cleansing – but was masked by an affirmation of progressive values. Specifically, in this case, the project of afforestation and of 'making the desert bloom' created a literal cover for the destruction of the native landscape, while seeming at face value to represent laudable – even, again, utopian – ecological values:

The act of affirming the positive value of greening the land is inextricably bound up with the negative reality of ethnic cleansing and ecocidal disappearance in such a way that emphasizing the former makes the latter fade away into a carefully managed void: the absence *is* the trees" (ibid: 35).

So central are such afforestation projects to the Israeli identity and its relation to the manufactured landscape it inhabits that, fifty years into the existence of the Israeli state, the defence of the forest surrounding Jerusalem from commercial development plans advanced by the municipality could take the form, not of a purely environmentalist initiative, but an expression of "a Zionism that has, as part of its territorial ethic, a concept of stewardship that favors the maintenance of the forest" (Cohen 2002: 210).

Writing about the same forests surrounding Jerusalem, Makdisi notes how the pine forests overlooked by Yad Vashem, the World Holocaust Remembrance Centre, hide the ruins of Deir Yassin, site of the 1948 massacre in which at least 107 Palestinian villagers were killed by Zionist paramilitaries (2022: x). Similarly, Davis writes about venturing into the Martyrs' Forest (Ya'ar HaKdoshim), one of the forests established by the JNF, to locate traces of the Palestinian villages of Suba and Bayt Mahsir, depopulated and destroyed in the Nakba. As she writes, "the geography is such that without knowledge of the Palestinian villages' existence in the past it would have been impossible to know what they were once here" (2011: 2). Ilan Pappe – who explicitly refers to the project as an act of memoricide – writes that "the true mission of the JNF has been to conceal these visible remnants of Palestine not only by the trees it has planted over them, but also by the narratives it has created to deny their existence" (2006: 228).

Such a far-reaching praxis of erasure requires considerable and ongoing effort. Moor writes that "the cleansing of Zionist memories has been a high-maintenance and continuous undertaking which has required diligence and an entire nation's energies" (2024: 16). Inevitably, even with the resources and efforts placed in the service of this agenda, the success of such an attempt to rewrite history and remake the landscape can never be total. Pappe writes that "the year 1948 is not a distant memory, and the crimes are still visible in the landscape for the present generation to behold and understand" (2024: 24).

It is unsurprising that there have been attempts to counter this narrative by preserving, and disseminating, the remaining traces of pre-1948 Palestinian life in the territories claimed by Israel. Perhaps the most important such work is Walid Khalidi's *All that Remains,* which, through extensive field and historical work, documents the 418 villages which were depopulated in the Nakba of 1948, the vast majority of which "have literally been wiped off the face of the earth" to the point that they "might as well never have existed" (1992: xxxii). One of the main ways in which Khalidi's book attempts to counter this erasure is through the presentation of photographs of the villages' surviving ruins. The book's front cover, for example, features a photograph of the ruins of the village of Suba, west of Jerusalem, whose history can be traced back to the Persian or Hellenistic period, and which had a population of 620 before it was depopulated during the Nakba (ibid: 318). The presentation of the ruin-image becomes a political act, a counter to the attempt at memoricide inherent in the Israeli clearing of the ruins. It says *we were here,* insisting on a history which – especially upon the book's publication in 1992, but, as Pappe notes, even now, more than thirty years later – is still in living memory.

However, Khalidi points out the difficulty inherent to the endeavour – in most cases, there is nothing left to photograph:

> The photographs have been selected to 'show' something – the remains of houses, public structures, current uses of what remains. In this sense, they are not representative, as the vast bulk of the photographs show largely empty sites. (ibid: xxiv)

Of the village of al-Ghazzawiya in the Jordan Valley, whose population in 1944/5 stood at 1,640, he writes – by way of example – "No physical evidence indicates that the village ever existed; the entire area has been levelled and is now cultivated by Israeli farmers" (ibid: 49).

Even where traces of the lost villages do remain, one has to look carefully to find them. Of the village of Kuwaykat, for example, Khalidi writes the following:

> Little remains of the village except the deserted cemetery, completely overgrown with weeds, and rubble from houses. Inscriptions on two of the graves identify one as that of Hamad 'Isa al-Hajj, and another as that of Shaykh Salih Iskandar, who died in 1940. The shrine of Shaykh Abu Muhammad al-Qurayshi still stands but its stone pedestal is badly cracked. A forest of pine and eucalyptus has been planted on the site. (ibid: 22)

Similarly, writing of the village of Saffuriyya, north of Nazareth, which housed 4,330 residents in 1944/5, he notes:

> Only a few houses remain on this site, including those of 'Abd al-Majid Sulayman and 'AmiMawjuda. Otherwise the site is covered by a pine forest planted by the Jewish National Fund to commemorate a number of persons and occasions [...] Zahir Umar's fortress still stands atop the hill, though some of its walls have collapsed. It is ringed by excavation sites. (ibid: 353).

As Makdisi writes about this same site, "the densely planted pine trees clustered together on a hilltop look suspiciously and unnaturally out of place because they are – they are covering up the ruins of an entire town" (2022: 26). In this way, the active clearing or concealment of Palestinian ruins from the contemporary landscape of Israel has been one of the key ways in which "Zionism has hidden, or caused to disappear, the literal historical ground of its growth" (Said 1979: 57).

The ruin-images of Gaza

Against this background – decades of Israeli efforts to actively destroy, erase or cover up the traces of Palestinian ruins and deny their existence, in the attempt to fabricate the myth of a 'land without a people' as an unclaimed wilderness ready for the foundation of a Zionist state – the images emerging from Gaza over the past two years make for a stark contrast.

Though they are, in significant ways, products of the same discourses of political and economic power, a revealing comparison can be drawn between the Harey Zahav image and the AI-generated video of "Trump Gaza" shared by Donald Trump on the Truth Social network in February 2025 (Krever and Salem 2025). The video does start with familiar images of the ruins of Gaza – families walking through streets of rubble between shattered apartment blocks – but only as a decontextualized rendition of the familiar tropes of poverty, violence, deprivation and neglect still associated, in a continually perpetuated Orientalism, with the Middle East; a 'Before' to contrast with the 'After' that forms the bulk of the video. Here, what we see is a sequence of images of a generic tropical seaside resort – high-rise towers behind palm-lined beaches, superyachts anchored in azure waters, sports cars driving down lavish streets, attractive women dancing in nightclubs, an ostentatious golden statue of Trump, Elon Musk throwing dollar bills in the air at a beach party, and a minaret or two for local colour. A utopia of international neoliberal wealth, if one is that way inclined, and one that is emphatically the no-place that implies: so non-specific is the generalized image of ostentatious luxury, enabled by the algorithmic bias towards the hegemonic norm of image generation software, that the scene could just as well be Miami or Dubai, Cozumel or Benidorm.

Like the promise held out by the Harey Zahav image, this utopia in the language of real estate and commercial development implies the levelling of the cities, towns and villages that currently occupy the territory, and the ethnic cleansing – whether through displacement or extermination – of its 2.1 million Palestinian inhabitants. However, in the video, this foundational violence remains only implicit. The mechanisms of "forced forgetting" are already in operation – the redevelopment plan is presented as a solution to the current ruined state of Gaza, distracting attention away from the question of how Gaza was reduced to ruins in the first place.

Instead, in the Harey Zahav image, the violence is inescapable. There is no "forced forgetting," no pretense of a blank slate awaiting development – on the contrary, the image draws specific attention to the intertwining of the proposed luxury real estate development and the ethnic cleansing which is its precondition. The ruin and what will replace it stand side by side, in explicit causal relation: the luxury beachfront villas, as potential settlement, are the reason for the ruins.

Two caveats need to be made here. First, it is not the violence itself that is new. By many measures – number of deaths per day, number of children killed, percentage of housing stock destroyed, etc. – the current assault on Gaza constitutes one of the worst crimes against humanity in recent history (Marketic 2025) – but what it marks is a drastic intensification of policies of ethnic cleansing that have already been directed towards Palestinians by the Israeli state for decades. The grand imperial projects that were the political products of Enlightenment humanism were the cause of countless wars and genocides (and produced endless fields of ruins) over the past centuries – as we have seen, all the way into the great modernist architectural projects of the twentieth century, and into the establishment of the state of Israel as a settler-colonial project founded on ethnic cleansing, or, as Pappe put it, "a nineteenth-century colonialist project extended into the twenty-first century" (2015: 22). However, what has changed is that this violence is now out in the open – it is celebrated for its own sake, rather than swept under the carpet or concealed behind the banners of civilisation, progress, development, and 'human rights."

Second, as with all blanket statements, a degree of nuance is in order. The fact that Israel has killed at least 192 journalists as of August 12, 2025 – the US-based NGO Committee to Protect Journalists has called it "the deadliest period for journalists since CPJ began gathering data in 1992" (2025) – indicates an ongoing attempt to control the media narrative and to conceal the extent of the atrocities being committed in Gaza, as do the routine attempts to offer justifications when specific atrocities are revealed. Nonetheless, ruin-images like the Harey Zahav image or the IDF soldiers' Gaza selfies are not isolated cases, and can be understood in the context of a shift in Israeli – and, arguably, in a broader sense, Western – political discourse.

The Israeli journalist Gideon Levy has referred to this as a "legitimization of evil," which he sees as having taken hold across much of the Israeli political establishment: "it's not only the extreme right who have tainted the political discourse. The center also wants more blood, destruction, epidemics, and hunger, and is unashamed to say so openly" (2023). Examples of such discourses abound – Levy quotes Brigadier General Dado Bar Kalifa saying, in a televised April 19, 2024 interview, "In Gaza we were a herd of elephants that left ruins behind it," before noting that even this is not enough for the increasingly prevalent Israeli right represented by Itamar Ben-Gvir. The general tone of Israeli political discourse, Levy argues, takes a similar tone: "We will strike and ravage, ruin and plunder [...] everything by force, without limit" (2024).

Perhaps, as Antony Loewenstein writes, this is a consequence of the awareness that the media myths that Israel had long succeeded in perpetuating – the idea of Israel as a "liberal oasis" in the Middle East, a bastion of democracy and human rights, and as the victim of unprovoked antisemitic violence – had unravelled and no longer held sway among many in the West, including young Jewish Americans (2024). In 1992, Walid Khalidi could already say that "the colonization of the homeland of the Palestinians took place in the modern age of communication and continues in full vigor under the glare, however fitful, of the electronic mass media" (1992: xxxi). The media landscape has changed considerably since then, but, in the context of social media and always-on digital channels of communication, Khalidi's statement is even more apposite now, as we witness the atrocities in Gaza unfold on our phones day after day.

Alternatively, these images could take their place within a broader discursive shift in Western politics, where the wide-ranging slide towards the far-right has increasingly mainstreamed and legitimized openly racist, xenophobic, sexist and otherwise exclusionary and violent language and policy. As the journalist Adam Serwer has said of the first Trump administration – and which is even more true of the second – "the cruelty is the point" rather than a by-product (2021); it is the centrepiece of the political theatre and a motivator of policy in a political climate where the idea of shared, communal hope no longer holds even as a rhetorical strategy. In the Palestinian context, Abdaljawad Omar speaks of a "theater of humiliation" that has been performed by the Israeli state with particular intensity in the years since October 7 (2025).

Writing about the Gaza crisis, the historian and genocide scholar Taner Akçam remarks precisely upon this shift in the way in which power mediates its own violence:

Violence is no longer hidden. It is no longer even regretted. Political figures have not merely tolerated this destruction; they have begun to brag about it. [...] Ethnic cleansing is rebranded as real estate opportunity. There is no shame in it. Violence has become a branding strategy. (2025)

The violence that power has always wielded – in Israel and elsewhere – has tended to be concealed by the "politics of denial" (Moor 2024: 13), masked behind the affirmation of progressive values like sustainability, democracy, diversity, and tolerance (Makdisi 2022), especially in the case of a state like Israel that closely aligns itself with the Western liberal order. The atrocities in Gaza since October 7, 2023 show, conclusively, that this is no longer the case: violence and the exercise of state power are not concealed but celebrated.

For Akçam, this radical shift in the way military, political and cultural power engages with the mediated representation of violence – and the ruins it leaves behind – represents a "civilizational rupture" (ibid.). He draws on Hannah Arendt's argument, in *Between Past and Future,* that, following the Holocaust, the West found itself in the condition of the "end of tradition," as the established moral categories of Enlightenment humanism and classical justice could no longer hold (2006). Akçam argues that the unfolding genocide in Gaza represent a similar crisis for the post-war liberal order:

The civilizational claim modernity made about itself was rooted in the Enlightenment — a project that presented itself as a turning point: a moment when humanity would restrain its violent instincts through reason, law, and universal principles. That promise, briefly renewed after 1945, took form in a new moral grammar — an extension of the ethical corpus shaped in the aftermath of the Holocaust, which was seen as a civilizational break. What we are experiencing now is the end of that tradition. (2025)

The Harey Zahav image – whether it was, in fact, intended as a joke or as a serious proposal – thereby casts a harsh light upon the conditions much of the world's population already live in, and those of us who are privileged enough not to for the moment can dread. It presents a world reduced to ruin by the endurance of colonialist exploitation, the re-entrenchment of authoritarian power, the neoliberal military-industrial complex, the rise of fascism, the revelation of the hollowness of the so-called liberal world order, the climate crisis and environmental catastrophe, masked by the increasingly vacuous promise of digitally-rendered virtual utopias and displays of ostentatious wealth within the gated communities of the global elite. Perhaps it is in that casting of the light itself – in showing us the true face of the twenty-first century – that these ruin-images can engender a form of hope: a hope that the horror and the anger they engender can be productive, and instil political change.

References

Abayov, Tsafrir (2024): "How this AP Photographer Caught this Image of Israeli Soldiers Taking a Selfie at the Gaza Border." In *The Independent*, March 3, 2024. https://www.independent.co.uk/news/gaza-ap-israeli-ashkelon-people-b2506223.html Last visit: August 17, 2025

Akçam, Taner (2025): "Thinking in Dark Times: The End of Tradition and the Crisis of Holocaust and Genocide Studies." *The Armenian Mirror-Spectator*, April 9, 2025. https://mirrorspectator.com/2025/04/09/thinking-in-dark-times-the-end-of-tradition-and-the-crisis-of-holocaust-and-genocide-studies/ Last visit: August 17, 2025

Al Kateab, Waab/Edward Watts (2019): *For Sama*. Republic Film Distribution. [Film].

Alvarez, Alez (2014): *Native America and the Question of Genocide*. Lanham, MD: Rowman & Littlefield.

Arendt, Hannah (2006): *Between Past and Future: Eight Exercises in Political Thought*. London, UK: Penguin.

Azaryahu, Maoz (2004): ""A Brief and Concentrated Image of Our Ambitions and Problems": The Israeli Pavilion at the World's Fair, Brussels 1958." In *Israel: Studies in Zionism and the State of Israel: History, Culture and Society*, 6, pp. 1-30.

Benjamin, Walter (2003) [1940]: *Selected Writings, 1938-40*, H. Eiland and M.W. Jennings (eds.). Boston, MA: Harvard University Press.

Bier, Jess (2017): *Mapping Israel, Mapping Palestine: How Occupied Landscapes Shape Scientific Knowledge*. Cambridge, MA: MIT Press.

Bloch, Ernst (1986) [1954]: *The Principle of Hope*, N. Plaice, S. Plaice, and P. Knight (trans.). Oxford, UK: Blackwell.

Chernov, Mystyslav (2023): *20 Days in Mariupol*. PBS Distribution. [Film].

Cohen, Shaul E. (2002): "As a City Besieged: Place, Zionism, and the Deforestation of Jerusalem." In *Environment and Planning D: Society and Space*, 20, pp. 209-230.

Committee to Protect Journalists. (2025): "Journalist Casualties in the Israel-Gaza War." October 13, 2023 (updated August 12, 2025). https://cpj.org/2023/10/journalist-casualties-in-the-israel-gaza-conflict/ Last visit: August 17, 2025.

Davis, Rochelle A. (2011): *Palestinian Village Histories: Geographies of the Displaced.* Redwood City, CA: Stanford University Press.

Dayan, Linda (2023) "No, an Israeli Real Estate Company is not Selling Beachfront Homes in Gaza." *Haaretz*, December 20, 2023.

Eshel, Amir (2010): "Layered Time: Ruins as Shattered Past, Ruins as Hope in Israeli and German Landscapes and Literature." In Hell, J. and Schönle, A. (eds.), *Ruins of Modernity*. Durham, NC: Duke University Press, pp. 133-149

Fanon, Frantz (2001) [1961]: *The Wretched of the Earth*, Constance Farrington (trans.). London: Penguin.

Goethe, Johann Wolfgang von (1831): "America, You've Got it Better," Daniel Platt (trans.). Schiller Institute. https://archive.schillerinstitute.com/transl/trans_goethe.html#America Last accessed: August 17, 2025

Herwitz, Daniel (2010): "The Monument in Ruins." In Julia Hell and Andreas Schönle (eds.), *Ruins of Modernity*. Durham, NC: Duke University Press, pp. 232-242

Herzl, Theodor (1960) [1902]: *Old-New Land,* trans. L. Levinson. Jacksonville, FL: Bloch.

Jameson, Fredric (2005): *Archaeologies of the Future: The Desire Called Utopia and Other Science Fictions.* London, UK: Verso Books.

Khalidi, Muhammad Ali (2001): "Utopian Zionism or Zionist Proselytism? A Reading of Herzl's Altneuland." In *Journal of Palestine Studies*, 30, 4, pp. 55-67

Khalidi, Walid (ed.) (1992): *All that Remains: The Palestinian Villages Occupied and Depopulated by Israel in 1948*. Washington, DC: Institute for Palestinian Studies.

Kirk, Terry (2008): "The Image of King Vittorio Emanuele II and the Remaking of Rome." In *The Court Historian*, 13, 1, pp. 35–49.

Krever, Mark/Mostafa Salem (2025): "'Trump Gaza is Finally Here!': US President Promotes Gaza Plan in AI Video." *CNN*, February 26, 2025. https://edition.cnn.com/2025/02/26/world/trump-promotes-gaza-plan-ai-video-intl/index.html Last visited: August 17, 2025.

Lebow, Katherine (2013): *Unfinished Utopia: Nowa Huta, Stalinism, and Polish Society 1949-1956.* Ithaca, NY: Cornell University Press.

Levy, Gideon (2023): "The Legitimization of Evil Will Remain With Israel Long After the War in Gaza Ends." *Haaretz*, December 31, 2023

Levy, Gideon (2024): "Israel's Right-Wing Can Never Seem to Get Enough Death and Destruction." *Haaretz*, April 21, 2024

Loewenstein, Antony (2024): "Zionist Media Myths Unveiling." In Antony Loewenstein and Ahmed Moor (eds.), *After Zionism: One State for Israel and Palestine.* London, UK: Saqi Books, pp. 163-173

Löwy, Michael (2017) [1988]: *Redemption and Utopia: Jewish Libertarian Thought in Central Europe,* Hope Heaney (trans.). London, UK: Verso Books.

Madley, Benjamin (2015): "Reexamining the American Genocide Debate: Meaning, Historiography, and New Methods". In *The American Historical Review*, 120, 1, pp. 98-139.

Makdisi, Saree (2022): *Tolerance is a Wasteland: Palestine and the Culture of Denial*. Oakland, CA: University of California Press.

Marketic, Branko (2025): "Israel's War in Gaza is One of History's Worst Crimes Ever." In *Jacobin*, August 5, 2025. https://jacobin.com/2025/08/israel-gaza-worst-crimes-ever Last visit: August 17, 2025

Miéville, China (2016): "Introduction." In *Utopia*. London, UK: Verso Books, pp.

Moor, Ahmed (2024): "Presence, Memory and Denial." In Antony Loewenstein and Ahmed Moor (eds.), *After Zionism: One State for Israel and Palestine*. London, UK: Saqi Books, pp. 13-22

Notaro, Anna (2000): "Exhibiting the New Mussolinian City: Memories of Empire in the World Exhibition of Rome (EUR)". In *GeoJournal*, 51, pp. 15–22.

Omar, Abaljawad (2025): "Marwan Barghouti, Itamar Ben-Gvir, and the Israeli Need to Humiliate." In *Mondoweiss*, August 16, 2025. https://mondoweiss.net/2025/08/marwan-barghouti-itamar-ben-gvir-and-the-israeli-need-to-humiliate/ Last visit: August 17, 2025

Pappe, Ilan (2006): *The Ethnic Cleansing of Palestine*. London, UK: Oneworld.

Pappe, Ilan (2015). "The Old and New Conversations." In Noam Chomsky and Ilan Pappe, *On Palestine On Palestine*, F. Barat (ed.). London, UK: Penguin.

Pappe, Ilan (2024): "The State of Denial: The *Nakba* in the Israeli Zionist Landscape." In Antony Loewenstein and Ahmed Moor (eds.), *After Zionism: One State for Israel and Palestine*. London, UK: Saqi Books, pp. 23-42

Procházka, Martin (2012): *Ruins in the New World*. Prague, Czechia: Litteraria Pragensia.

Riegl, Alois (1982) [1903]: "The Modern Cult of Monuments: Its Character and its Origins." K. W. Foster & D. Ghirardo (trans.). In *Oppositions*, 25, pp. 21–51.

Robinson, Nicola (2013): "Utopian Zionist Development in Theodor Herzl's *Altneuland*." In *Green Letters: Studies in Ecocriticism*, 17, 3, pp. 223-235.

Royle, Nicholas (2003): *The Uncanny*. Manchester, UK: Manchester University Press.

Said, Edward (1979): *The Question of Palestine*. London, UK: Vintage.

Said, Edward (2003) [1978]: *Orientalism*. London, UK: Penguin.

Serwer, Adam (2021): *The Cruelty is the Point: The Past, Present and Future of Trump's America*. New York, NY: One World.

Settis, Salvatore (1997): "Introduction." In Michael S. Roth, Claire Lyons and Charles Merewether (eds.), *Irreversible Decay: Ruins Reclaimed*. Los Angeles, CA: The Getty Research Institute for the History of Art and the Humanities.

Shai, Aron (2006): "The Fate of Abandoned Arab Villages in Israel, 1965-1969." *History & Memory* 18, 2, pp. 86-106.

Stolow, Jeremy (1997): "Utopia and Geopolitics in Theodor Herzl's *Altneuland*." In *Utopian Studies* 8, 1, pp. 55-76.

Ruined Places

A Performance of Decay
Amitesh Grover's Site-Specific Theatre, The Money Opera

Rahul Bishnoi

Abstract

This article discusses Indian theatre director Amitesh Grover's experimental performance, The Money Opera *(2023 – ongoing), as a site-specific critique of urban and ecological decay in contemporary India. Staged within the ruins of abandoned buildings, the immersive production invites audiences to navigate a multilayered narrative space, encountering a diverse cast of characters – ranging from a young billionaire and a ghost to a poet and a stockbroker – who embody the contradictions and hidden truths of capitalist society. As audiences move through the decaying structure, their seemingly free choices are juxtaposed with the uncontrollable unfolding of events, mirroring the illusion of agency within capitalist systems. Through fragmented stories and songs exploring ambition, guilt, fear, desire, and community,* The Money Opera *exposes the pervasive fiction of money and the socioeconomic structures that perpetuate exploitation and inequality.*

Drawing on Walter Benjamin's concept of ruins as allegories of capitalist failure and Quetzil Castañeda's analysis of heritage-making, the essay argues that Grover's work creates a hauntological presence, reanimating suppressed histories and marginalized voices within the ruins. By employing non-actor performers and immersive audience engagement, the performance challenges the abstraction of space under capitalism and critiques sanitized narratives of progress and heritage. The essay situates The Money Opera *within the broader tradition of site-specific theatre, and points out its unique contribution to postcolonial contexts like Goa, where rapid urbanization and tourism have exacerbated ecological and social exploitation. Ultimately, the performance reimagines ruins as dynamic sites of memory and resistance, offering a powerful lens to interrogate the performativities of decay in neoliberal capitalism.*

Keywords

site-specific theatre, urban ruins, postcolonial performance, Amitesh Grover

> "... j'ai de chaque chose extrait la quitessence, Tu m'as donne ta boue et j'en ai fait de l'or
> [I extracted the quintessence from everything, you gave me your mud, and I made gold from it]"
>
> Baudelaire (2013: 180)

Introduction

The French poet Charles Baudelaire is argued to be one of the first poets who spoke of the urban experience – particularly, a life in Paris – in his poem *Le Cygne* (Laforgue 1903: 111-113). The epigraph illustrates the essence of urbanity as a provocative thought: a city perpetually resides in the ruins. Old structures continuously decay with the course of time, or are made to fall. New constructions replace them to become a part of the city. In this process that resembles the paradox of the Ship of Theseus[1], the city is simultaneously dismantled and rebuilt, preserving only the mud as its common denominator – the mud of decayed ruins and the concrete that cements the bricks of the future. Baudelaire's assertion, "you gave me your mud, and I made gold from it," captures the transformative power of art in modernity. Here, art becomes a catalyst, as T.S. Eliot[2] might describe it, converting the suffering and disarray of urban industrial life into an object of pleasure and reflection.

This transformative ethos serves as a departure point for this article to discuss Indian theatre director Amitesh Grover's experimental performance *The Money Opera* (2023-Ongoing). It is an immersive, site-specific work staged within the desolate remnants of abandoned buildings in different cities – Goa and Delhi, so far. The performance situates itself against the backdrop of urban and ecological decay created by the rapid real estate expansion in contemporary India. Through its site-specific design, *The Money Opera* subverts the proscenium-based theatre's formation of passive audiences that consume the served spectacle. Instead, the play invites the audience to participate in an entangled network of parallel performances happening within the decaying building. A variety of characters animate each floor, exposing capitalism's hidden, oppressive layers and the pervasive fiction of money. This essay aims to introduce *The Money Opera* as a

---

1   Referring to The Theseus' paradox, a thought experiment that raises the question of whether an object that has had all of its components replaced remains fundamentally the same object
2   T. S. Eliot, in his essay *Tradition and the Individual Talent* (1919), introduced the concept of the poet as a catalyst. He asserted that the poet's role in writing poetry is that of a catalyst in a chemical reaction. The poet does not merely express personal emotions in the act of writing but rather serves as a conduit for the collective cultural consciousness.

critical spectacle to rethink the performativities of decay, particularly within the context of ruins-as-tourist sites. Ruins, often viewed as remnants of the past, have historically been imbued with symbolic, aesthetic, and cultural meanings, serving as markers of historical grandeur, architectural decay, or collective loss. However, in the contemporary context, particularly under neoliberal capitalism, ruins are increasingly commodified as heritage, packaged to evoke nostalgia, awe, or even escapism. These spaces are frequently sanitised and reimagined, their histories curated to fit marketable narratives that obscure their socio-political and ecological implications. By situating the performance within the ruins of abandoned buildings, the paper begins a conversation to seek the potential of site-specific theatre to interrogate the violent processes of commodification and to resist the abstraction of space.

Synopsis

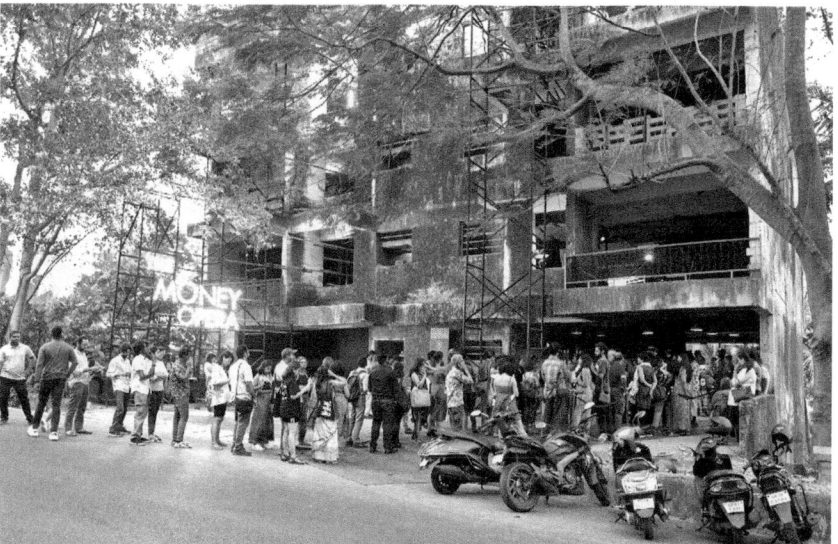

Fig. 1: The facade of the abandoned building – the performance site for The Money Opera (Agarwal 2022a)

In 2022, the debut performance of *The Money Opera* took place in Goa, one of the most popular tourist destinations of India. Amitesh Grover surveyed the abandoned sites and buildings in the region and selected a five-storey building that has been lying empty for about ten years (Nath 2022). In his own words, Grover described the building of half-built walls as "a site that is sinister in some ways, it carries the burden of an unfulfilled past. It is as if it is yearning for a future that is not there" (Nath 2022). In the performance, the uninhabited building becomes a metaphor for capitalist excess and unfulfilled potential. Grover and his group transformed the decaying building into a scenography-led multilayered performance site where each floor was associated with a dramatic storyworld of characters. Within this haunting venue, *The Money Opera's* cast of characters consisted of a young billionaire, a ghost, an inheritor, a child goddess, a thief, a poet, a stock-broker, a rebel, an explorer, and several others, some of whom produce the capitalist systems of oppression while the others survive, fight and struggle against it, both resulting in decay and ruins. The characters were performed by a cast comprising both seasoned actors and real-life professionals, telling supposedly unrelated solo-performed narratives in different locations of the abandoned building, with money being the common denominator across all. The audience were welcomed and introduced to the performance by 'the Landlord' figure in the play, and then set free in the open marketplace of stories to wander and fatefully encounter the parallel story arcs of the Money Opera.

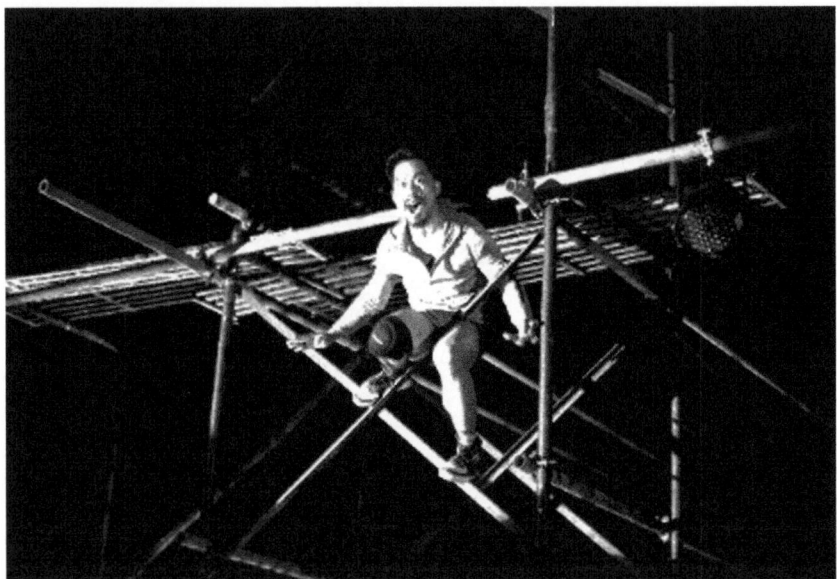

Fig. 2: Performance Still from The Money Opera, showcasing dramaturgies of disconnected narratives linked through the site (Agarwal 2022b)

The title *The Money Opera* has similarities with Bertolt Brecht's seminal work, *The Threepenny Opera* (1928), which itself was a radical reworking of John Gay's *The Beggar's Opera* (1728), although Grover does not explicitly says so in the prologue or in his interviews. The connection between these two works extends beyond titular similarity, as both operas interrogate the commodification of human life and the pervasive fiction of money. In *The Threepenny Opera*, Brecht's character Macheath famously declares, "What is the robbing of a bank compared to the founding of a bank?" (Brecht 1928: 78) – a line that encapsulates the systemic critique at the heart of the play. Likewise, *The Money Opera* stages a cacophony of voices – ranging from billionaires to labourers – who collectively expose the hidden layers of capitalist extraction and exploitation. In fact, Grover's use of parallel, disconnected narratives within the abandoned building can be seen as a 'site-specific' adaptation of Brecht's alienation effect[3]. By forcing the audience to navigate the decaying space and encounter disparate storylines, *The Money Opera* disrupts the passive consumption of spectacle and instead compels an active, critical engagement with the performance. This approach not only critiques the commodification of space and labour but also reimagines the role of the audience as co-participants in the act of meaning-making.

The term 'site specific' is quite commonly used for performances such as *The Money Opera* that venture outwards from the proscenium theatre spaces and into 'new' or 'found' spaces. Scholars like Joanne Tompkins [*Performing Site-Specific Theatre* (2012)], Nick Kaye [*Site-Specific Art: Performance, Place, Documentation* (2000)], and Bertie Ferdman [*Off Sites: Contemporary Performance beyond Site-Specific* (2018)] have thoroughly traced the development of this form. Site-specific theatre is argued to be a response to the homogenizing forces of global neoliberalism, which often reduce places to interchangeable commodities (Ferdman 2018). These performances facilitate a confrontation between the audience and the unique histories, ecologies, and social relations of a particular location. In doing so, they expose the violent processes of capitalist extraction that transform living, dynamic spaces into static, marketable assets. Scholars like Cathy Turner (2012) and Bertie Ferdman (2018) also acknowledged that in the majority of site specific performance studies, the emphasis has been on European artists, with particular focus on the United Kingdom (Ferdman 2018: 26). *The Money Opera*, therefore, offers an opportunity to explore theoretical contributions of what capitalist extraction means in a post-colonial space. It is more so relevant in the state of Goa which was liberated from Portuguese rule in 1961, fourteen years after the independence of India (Rao 1963).

What is the political significance of the place in this performance? Recent scholarship has attributed the site-oriented approaches of theatre artists post the 1980s to larger historical shifts in three areas – technology, the global economy, and ecological crises (Ferdman 2018). These life-altering changes demand a new way of thinking about making art, and, to varying degrees, have led to blending artforms, resisting categorisations, and the creation of work that in process determines its (new) form – all of which come together in the form of what has been generically referred to as site-specific performance. Contemporary approaches to

---

3   Bertolt Brecht's alienation effect, or *Verfremdungseffekt*, is a theatrical technique that aims to distance the audience from emotional involvement in a play. By reminding the audience of the artificiality of the performance, Brecht encourages critical thinking and analysis of the play's social and political messages, rather than emotional catharsis (Brecht 2013).

site specificity now question theatrical form by blurring the traditional fictional-versus-real distinction of experience and suspension of disbelief; mapping new practices for audience/actor landscapes; fabricating sites from physically bound, conceptually constructed, or virtual spaces; staging live situations in real/nonreal and often mediated encounters; and extending preconceived notions of time and space.

The site-specificity of the performance was rooted in the state politics of Goa. Being a popular tourist destination in India, Goa has become an epicentre of foreign and local investment leading to land acquisition for hospitality projects. *The Money Opera* invariantly aimed to platform the marginalised voices of the region as they suffered and engaged with these developing changes in the socio-ecological locality. Grover facilitates this through the casting of non-actor performers belonging to the region, such as Mamata Deepak Verlekar, a Konkani[4] poet who narrates anti-system poetry on corruption in land use and mining. Likewise, the cast included a stockbroker who handled the financial portfolios of the twenty richest Goan families and talked about witnessing the lives of wealthy people who did not understand money. The oceanographer character in the play, Dr. Aninda Mazumdar, performed the story of the now polluted sea, where mutated species survived on the toxic materials of industry. Therefore, *The Money Opera* sings an ensemble political poetry that comments on the performativity of Goa, which was mass marketed by the businesses as the tourist destination for your vacation or destination weddings. The play isolates the actors of this sociological network behind the tourist economy of Goa, and assembles them as a temporal collage in the disintegrating space of the building.

The thematic of decaying architectures and tourism is a well recognised and studied feature in the present neo-liberal economies. Latin American anthropologist Quetzil Castañeda (1996) investigates the Mayan ruins of Chichén Itzá in Mexico and argues that the concept of ruins does not exist in the language or the spatial sensibilities of the local residents. He discovered that Mayan people in Yucatán refer to the remains of the peninsula as *xlapak* ("old walls") and the remains of Spanish forts are called *fuertes* ("forts") (ibid). It is the archaeologists and Mayan[5] workmen who have selectively and strategically reorganised the decaying rubble of the Mayan city according to their own imaginings of the past, and constructed a fantasy of a unified object that exists in the form of ruins in the present (ibid). The repurposed remains of the past city are marketed as

---

4   Konkani is a regional language of Goa.
5   'Mayan workmen' refers to the industry employed Mayan people who work for archaeologists and/or tourism developers. They assist in reconstruction of the sites of ruins to align with external imaginings of the past. Castañeda argues that this process involves the strategic reshaping of the physical and symbolic landscape to create a marketable narrative of "ruins" as objects of heritage and tourism. (Castañeda 1996)

'heritage' and then transformed into tightly managed places where visitors pay to contemplate the relics that they are ordered to photograph but not to touch. The performativity of the decaying object is fetishised in the heritage-making process. Castañeda calls ruins an "authentic invention of modernity" as they are "the copy of an original that never existed" (ibid: 48–49).

Another way to look at the political ontology of ruin is through Theodor Adorno's logic of disintegration. In *Negative Dialectics*, Adorno underlines the critical potential of *negativity* to dismantle the illusion of completeness and coherence in the given reality (Adorno 2003 [1966]). This idea builds on Walter Benjamin's critique of capitalism's "bourgeois dream-world," where the spectacle of architectural and spatial forms serves as an ideological fantasy masking deeper contradictions (Benjamin 2015). For Benjamin, an unequivocal admirer of Baudelaire, decaying infrastructure symbolises the collapse of this fantasy, embodying destruction and disintegration. The 'mud' of the destructive ruins were allegories of critical negativity to both Adorno and Benjamin, challenging the reified positivity inherent in capitalist ideologies, such as those crystallised in 19th-century Parisian architecture. My analysis of *The Money Opera* draws from this Adornian and Benjaminian project but pushes it further to invite interrogations on the performative potential of the ruins. While 'the ruin' certainly evokes rupture, it also evokes a unified object that elite sensibilities often treat as a fetish that ought not be disturbed.

The Money Opera, in my view, is a reconfiguration of a decaying architectural space, except its traces do not go as far back as the Mayan civilization. Grover's performance does not seek to reconstruct an imagined or glorified history of the abandoned building. Instead, it foregrounds the process of ruination itself, presenting the site as a living archive of ecological, economic, and human destruction. As Gastón R Gordillo argues in his book *Rubble: The Afterlife of Destruction* (2014), commodification reduces the sensory texture of places to "quantifiable, homogeneous abstractions to be sold and bought" (ibid: 8). *The Money Opera*, at its core, critiques the abstraction of land as a purely monetary resource and hosts the forgotten and ghostly connections of the land with people, animals, water and other constituents of the lifeworld of Goa. If the ruin is thought as a part of the abstraction of space that is subject to ideological erasure in narratives that present it as 'heritage', then the performance, I argue, creates a hauntological presence of that erasure. Moreover, Benjamin's emphasis on detritus and garbage as sites of critical potential resonates with the performance's engagement with ecological decay. The polluted sea, mutated species, and toxic industrial waste referenced in the play are not merely background details but central elements of its critique. These remnants of capitalist exploitation, often hidden or ignored in dominant narratives, are brought to the forefront, challenging the audience to reckon with the ecological devastation wrought by unchecked development. In this way, *The Money Opera* extends Benjamin's critique of the phantasmagoria to the contemporary context of neoliberal capitalism, where the abstraction of space and the commodification of nature have reached new extremes. The re-colonised postcolonial land, once teeming with biodiversity and ecological significance, is stripped of its natural and cultural value to be transformed into a concrete edifice – a speculative commodity that prioritises financial gain over sustainability.

The performance also attends to the hidden human costs embedded within construction of luxurious hospitable spaces such as resorts and hotels in Goa[6]. Grover is aware of portraying the "sighs of exhausted labour" (Nath 2022) that underpin the construction of urban towers designed to serve industries such as tourism. These places, often celebrated as markers of progress, conceal the sacrifices of the labourers whose bodies and livelihoods are consumed in the process. The structure, once envisioned as a monument to economic prosperity, is abandoned, becoming a stark reminder of the ideological apathy inherent in capitalism's refusal to serve any purpose beyond profit. The abandoned building thus embodies the ghosts of those it has consumed – both human and ecological. The performance positions the ruin as a repository of stories and traumas, haunted by the memories of the lives and landscapes sacrificed in its making. Drawing on the work of theorists such as Tim Edensor (2005), who argues that ruins are spaces where traces of the past are reactivated and reimagined, *The Money Opera* irks its audience to engage with the building as a site of both memory and critique. The incomplete walls and hollowed-out floors resonate with the silences and absences left in the wake of capitalist exploitation, rendering the ruin a symbolic marker of systemic failure.

Grover's performance functions as both a critique and a meditation. By situating *The Money Opera* within the ruin, the production does not merely narrate a story of decay; it transforms the ruin into a performative space that critiques the socio-economic structures responsible for its condition. The performance becomes a rumination on ruination itself, insisting that the engine of claiming and terraforming a place into a money-making enterprise runs out of steam, eventually and inevitably. As Ann Laura Stoler (2013) notes, ruins are not just remnants of the past but active, ongoing processes that expose the inequalities and violences of the systems that produce them. In *The Money Opera*, the ruin is not a static object of nostalgia but a dynamic, living critique – a visceral confrontation with the remnants of failed ideologies and unsustainable practices. In doing so, *The*

---

6   At the time of writing this article, *The Money Opera* has only been staged in two locations (Delhi and Goa). As the performance travels to other sites across India, it will be fascinating to observe how it engages with diverse local narratives of decay and ruination. For instance, in the Andaman and Nicobar Islands, the performance could inhabit the ruins of British-era prisons, which stand as haunting reminders of colonial oppression; in Jharkhand and other parts of eastern India, it might explore the ruins of indigenous tribal villages displaced by large-scale dam projects; in Kerala, the performance could confront the ruins left by climate change-induced floods and landslides, to name a few possibilities.

Money Opera expands the role of site-specific performance, engaging with the ruin not only as a physical space but as a complex symbol of ecological devastation, labour exploitation, and ideological apathy.

References

Adorno, T. (2003) [1966]: *Negative dialectics*. London: Routledge.

Agarwal, R. (2022a): *The Money Opera [The facade of the abandoned building]*. Goa, India

Agarwal, R. (2022b): *The Money Opera [Performance Stills]*. Goa, India

Baudelaire, C. (2013 [1857]): *Flowers of Evil and Other Works: A Dual-Language Book*. Massachusetts: Courier Corporation.

Benjamin, W. (2015): *Walter Benjamin's Archive: images, texts, signs*. New York: Verso books.

Birch, A, and Tompkins, J. eds. (2012): *Performing Site-Specific Theatre: Politics, Place, Practice*. New York: Palgrave Macmillan.

Brecht, B. (1976): The Threepenny Opera. New York: Random House Inc

Brecht, B. (2013): "Short description of a new technique of acting which produces an alienation effect". The Twentieth Century Performance Reader, pp. 101-112

Castañeda, Q. (1996): *In The Museum Of Maya Culture*. Minneapolis: University of Minnesota Press.

Edensor, T. (2005): "The ghosts of industrial ruins: ordering and disordering memory in excessive space". *Environment And Planning D: Society And Space*, 23(6), pp.829-849.

Eliot, T. S. (1975 [1919]): *Tradition And The Individual Talent*. Selected essays, pp.13-22.

Ferdman, B. (2018): *Off Sites: Contemporary Performance Beyond Site-Specific*. SIU Press.

Gordillo, G. R. (2014): *Rubble: The Afterlife Of Destruction*. Duke University Press.

Grover, A. (Director). (2022): *The Money Opera* [Theatre performance]. Goa, India

Kaye, N. (2000): *Site-Specific Art: Performance, Place, and Documentation*. London: Routledge.

Laforgue, J. (1903): *Mélanges posthumes (Vol. 4)*. Mercure de France.

Nath, D. (2022, December 21): *In An Abandoned Building In Goa, Amitesh Grover Creates A Layered Piece On Capitalism, Titled The Money Opera*. Indian Express. https://indianexpress.com/article/lifestyle/art-and-culture/amitesh-grover-the-money-opera-serendipity-arts-festival-8336597/

Rao, R. P. (1963): *Portuguese Rule in Goa, 1510–1961*. London: Asia Publishing House.

Stoler, A. L. (ed.). (2013): *Imperial Debris: On Ruins And Ruination*. Duke University Press.

Turner, C. (2012): Site-Specific Performance. Contemporary Theatre Review, 22(3), pp.425–426. https://doi.org/10.1080/10486801.2012.697730

Paratextual anarchaeology
Revisiting *WildStar* through its residual traces

James Manning and Lawrence May

Abstract

Videogames constantly fall into states of ruin as they, or the platforms and consoles they are accessed through, become inaccessible for different technical or commercial reasons. Conventional approaches to videogame preservation consider this a technical problem, proposing software emulation as the solution to revisit videogames but overlooking the significance of the player in co-constituting videogame experiences. Paratextual remnants offer the potential to reconstruct otherwise 'lost' and unplayable videogames, and emphasise the situatedness and multiplicity of play.

We propose a methodological intervention that allows situated encounters of play to be reconstructed from the past by drawing on digital ephemera created by players. We adapt the approaches of media anarchaeology (Zielinski 2006) and platform anarchaeology (Apperley/Parikka 2018) to videogames, to outline a paratextual anarchaeology. This modified method draws on player-produced paratexts and the feelings and sensations of play they invoke. We apply our 'paratextual anarchaeology' to revisit WildStar *(Carbine Studios 2014), a short-lived PC-based MMORPG, and explore three renditions of the game, characterised by player feelings of* warmth, belonging *and* melancholy. *In doing so, we illustrate the distinctive contribution this method offers for constructing speculative, manifold accounts of past videogame experiences.*

Keywords

paratexts, anarchaeology, WildStar, affect, un/anarchives

Introduction

The ruins of increasing numbers of failed, forgotten and abandoned videogames collect all around us, rendered inoperable as services, servers, and circuitry are retired, discontinued, and fall into obsolescence. Conventional approaches to videogame preservation consider this a technical problem, proposing software emulation as the solution to revisit videogames. However, focusing solely on reproducing the playable artefact elides the significance of the player in co-constituting videogame experiences (Newman 2012; Giddings/Kennedy 2008; Morris 2003).

The methodological intervention we propose in this article allows the reconstruction of situated encounters of play from the past by connecting in more vernacular ways to alternative modes of remembrance driven by players. We modify the approach of "platform anarchaeology" (Apperley/Parikka 2018) – a method for constructing speculative media histories of failed, experimental, and long-vanished platforms derived from earlier media anarchaeological approaches (Zielinski 2006) and adapt it for videogames. We integrate the gathering and analysis of player-produced paratexts (reviews, discussion posts, gameplay videos, stories, etc.), embracing the capacity of these ancillary texts to capture and manifest emergent player experiences (May 2021).

The concept of the 'paratext' originates with literary theorist Gérard Genette (1997), but has latterly become inflected by burgeoning cultures of media convergence and audience (inter)activity and creativity (Gray 2010). Game studies scholars in particular have drawn attention to the capacity of paratexts to (re)define the meaning of games and play experiences for players (Consalvo 2007) and to capture experiences of individual and collective play (which are otherwise dynamic and fleeting) for others to witness (Mukherjee 2015). Paratexts are today a kind of digital ephemera, with officially sanctioned and user-created paratexts that reflect videogames and play experiences swirling together in a cacophony of media shared and consumed across a variety of platforms.

We apply our 'paratextual anarchaeology' to such digital ephemera to revisit *WildStar* (Carbine Studios 2014), a short-lived PC-based Massively Multiplayer Online Roleplaying Game (MMORPG) that ran between 2014 and 2018. Our case study enables us to demonstrate how the residual traces of play found in paratexts allow for reconstructions of a game that no longer exists. The distinctive contribution of this method lies in uncovering feelings, sensations, and situated experiences surrounding videogame texts often overlooked by formal archival methods and absent from institutional archives.

Other scholars have explored the 'ruins' of MMOs in ways complementary to ours. For instance, ethnographic approaches allow Mia Consalvo and Jason Begy (2015) to trace player sentiments throughout the full life cycle of *Faunasphere*, a short-lived MMO. Similarly, Celia Pearce (2009) turns to study the diaspora of players attempting to retain a sense of community after the official servers of *Uru: Ages Beyond Myst* close down. Justyna Janik's (2018) autoethnographic approach provides an interesting counterpoint and case study engaging more directly with the aesthetics of exploring firsthand the abandoned MMO *Meridian 59*.

Each approach raises interesting questions about the nature of playable worlds as they transition from occupied and fully functional to abandoned "ruin-like structures" (Janik 2018). Indeed, despite *WildStar* no longer being officially maintained by Carbine Studios, community-run servers still operate, providing a means to play (a reconstruction of) *WildStar* (see Belghast 2022). However, our intention here is to forego encountering the game for ourselves, to instead explore the potential to reconstruct *WildStar* from its paratextual remnants. Just as Janik is interested in the aesthetic encounter, we too are interested in how encounters of playing a game are expressed, and feelings and sensations are captured, within the paratexts that remain. Through such an approach we also ask what is lost when videogame preservation methods prioritise maintaining playable artefacts over the experiences they engender.

Preservation

This case study offers a timely exploration into the process of reconstructing a 'lost' game from its paratextual remnants. Memory institutions, such as galleries, libraries, archives and museums, operate as "part of an object-centred culture" (Swade 2002: 227). Built to resist the effects of change, traditional preservation strategies take for granted that an object's physical and material properties are fixed and self-evident. Yet, as Kenneth Thibodeau (2012) notes, traditional memory institutions find the polymorphic, shapeshifting characteristics of digital objects troubling, thus creating "an inherent tension between digital information, which does not stay still, and digital preservation, which quintessentially seeks to keep things in place, without significant change" (2012: 15). The apparent solution to this problem lies in adopting a digital preservation strategy that maintains operable, "functionally intact" software and systems (Swade 2002: 232). If the purpose of videogame preservation is to maintain *playable* artefacts, then software emulation provides a promising solution (Guttenbrunner et al. 2010; Pinchbeck et al. 2009).

Yet attempts to preserve technological artefacts alone risk eliding the significance of their use contexts (Winget 2011), and how different contexts modify the qualities of videogames (Guins 2014). Hence, the promise proposed by software emulation to retain access to a playable version (or versions) of a game appears inadequate: the practice is unable to account for how videogames are encountered, how they become embedded within certain sociohistorical contexts, and above all, how videogames are actually being played. To extract a videogame from its social, cultural and temporal context is to disregard the significance of its players in *cocreating* the game (Banks/Potts 2010; Potts et al. 2008; Morris 2003) or *co-constituting* the gameplay experience (Giddings/Kennedy 2008; Dovey/Kennedy 2006).

Such concerns lead Henry Lowood (2002) to urge archivists to extend their archival practices to include evidence of *how* videogames are performed and enacted within everyday use contexts. These "performance archives" challenge the notion that software emulation alone can save the future history of videogames (Lowood 2002: 15). A sentiment embraced by James Newman (2012: 38; original emphasis), who highlights the significance of documenting videogames "*in* and *at* play", moving the goal away from *game* preservation towards "game*play* preservation" instead. Dany Guay-Bélanger and colleagues (2023) have gone as far as to suggest that *without* gameplay preservation, there can be no gameplay studies at all. Given these perspectives, it appears difficult to imagine a future *without* access to the edifying records of player-based and player-created accounts of gameplay.

The capacity for player-produced paratexts to capture erudite descriptions of embodied play experiences has not been overlooked, however. Lowood (2006; 2011), for example, further explores performance archives through the loci of gameplay demos and machinima. James Newman (2011: 111) celebrates the capacity of player-produced walkthroughs to "capture and communicate the important qualities of games, as defined and understood by their players", while Niklas Nylund (2015) celebrates the idiosyncratic portrayals of gameplay captured in Let's Play videos. Derek Murphy (2015: 182) addresses narrative accounts of play more generally, finding a distinctive strength in their presentation of the "subjective and emotional experience" of videogame play. Such approaches align with what Austin Walker (2019) calls for in the formation of archives of *feelings*, recognising the "need to prioritize the archiving of contemporary, phenomenological, and anthropological records of play", including blogs, diaries, forum posts, fanart, user-created videos, text messages and more.

What begins to emerge across these accounts is that, and quite obviously, the degree to which something has been documented influences the type and extent of archival research that can later be performed on it. Less evident is that when documentation is lacking, retrospective recollections of past events can suffer from volunteerism, where contributors have a vested interest in how history might remember them. Recollections can be overly nostalgic (Scully-Blaker 2023: 514) and conform to gendered stereotypes (Kocurek 2015: 184). As such, Carl Therrien (2015; emphasis added) suggests that "voluntary witnesses must be confronted with *involuntary witnesses* [...] in order to properly document, explain and format history into a proper narrative". This desire to temper history, to seek out involuntary witnesses, undergirds our interrogation of paratextual materials.

Videogame artefacts maintained within formal institutional structures (such as the United Kingdom's National Videogame Museum and The Strong Museum of Play in Rochester, New York) alongside records sustained through continual community-driven efforts (such as the Internet Archive or the non-profit Video Game History Foundation) have already provided cultural historians opportunity to explore videogaming's past through paratextual analysis. Examples include recourse to British gaming magazines (Kirkpatrick 2012), *Softalk* magazines' readers' letters (Nooney/Driscoll/Allen 2020), and the 'invisible' histories residing in lifestyle magazines (Harkin 2024). What each of these cases makes clear is how paratexts capture a plurality of voices: something we look to retain within our proposed anarchaeological method. It also becomes clear that paratexts persist both inside and outside of archival settings. The resilience of digital records within the more quotidian, "living" archives (Carlin/Vaughan 2015) maintained by users of Reddit and YouTube provides, as we describe below, fertile ground for us to explore.

Over the last half-century, our understanding of the role archives perform within society has undergone significant change due, in part, to the widespread proliferation of digital technologies (Manoff 2004). For Abigail De Kosnik (2016: 35), in the formation and circulation of our "archives of mass media" cultural memory has gone "rogue". A practical way to approach this is to consider the role archivists play in archiving materials. For De Kosnik (2016: 18–9), fan sites are of particular interest noting that fan communities "have a long history of archive building" which makes them "exemplary digital memory workers at the level of content production as well as content preservation". Replacing outmoded views of archives as impartial carriers of information, De Kosnik (2016: 34) goes on to stress how each encounter *reproduces* the archive, giving shape and meaning to the events they record (see also Manning 2017). Federico Giordano and Bernard Perron (2014) similarly identify the characteristics of digital archives as being intangible, variable or fluid, and largely run by their users. Such fluidity displaces traditional views of the archive as authoritative, fixed and immutable, and extends understandings of what can be preserved, who does it and how this is achieved. To this end, Giordano and Perron's (2014: 20) alternative concept of the "anarchive" describes a *living* entity, refigured and reassembled within everyday use practices.

Paratexts

Embracing preservation studies' necessary move toward these *living* archives, we experiment with a methodology that draws centrally on the paratexts that circulate around videogames and player communities. This is in an effort to close the gap between the types of game materials that are typically formally archived and the situated experiences of play often lost along the way. Of course, paratexts are not necessarily created as part of conscious or deliberate efforts to archive or preserve games. There is such variety to the motivations behind, and the imagined 'purposes' of paratexts that we must, as Jan Švelch (2020) advises, be aware that while paratexts are grounded within specific socio-historical and cultural contexts, they can also exhibit different qualities to different audiences at different times.

Understood at their most fundamental level, however, paratexts created by the players of videogames capture, and manifest for those who later consume them, experiences of play. As Souvik Mukherjee (2015: 103) observes, play is a complex "multitelic experience", defined by player experiences of the videogame text and its stories that can be markedly different between different players, contexts of play, and individual play sessions. Naturally, all forms of media imbricate their audience in the process of bringing a work into being, but what principally distinguishes encounters with videogames from other forms of storytelling, Mukherjee argues, is their immanent combination of the experiential and the ephemeral (2015: 104). The production of a paratext means that those fleeting, situated moments can, in some form (whether as writing, an image, a video, art, or something else altogether) be captured.

While narrative (see Maclean 1991), fandom studies (see Barker 2017) and game studies (see Švelch 2020) scholars alike have attempted to demarcate and systematically differentiate the 'text' from 'paratext', our view is that the two occupy an equal footing. We therefore align ourselves with arguments that hold that 'paratext' is most usefully understood as a conceptual framework for addressing the potent *functional* potential of different types of texts (Birke/Christ 2013: 67) rather than ascribing hierarchical textual labels. Videogame 'texts' and the 'paratexts' created by players simply offer, in different ways, access to and participation in a play experience. Mukherjee (2015: 119) summarises this well, noting that it is not productive to think of a videogame's 'story' (and the experience of its play) as a singular, discrete entity on the one hand, and the paratext as another discrete entity simply created in reference to a source text. Rather, he explains, they have always been intrinsically entangled.

The paratext, then, is not seen as something subordinate to a 'source' text, but rather a moment of play that has been broken off from a wider, ephemeral encounter with a videogame. As such, paratexts exist as part of dynamic networks of meaning and memory discursively linking together players, texts and experiences (Webber 2023: 85), and allow us to zoom into (or out from) situated play encounters with play captured by players. The paratext also carries with it a kind of vitality that marks it as both a representation, and a manifestation of, the original play experience, evoking rich sensations of the emergent play experiences of the user who has created it (May 2021: 7–8).

Our proposed method presupposes that paratextual remnants are approached as though they are the 'ruins' of past videogame experiences. In order to interpret these traces as ruins, Robert Ginsberg's (2004) rumination on the aesthetics of ruins proves instructive. Ginsberg (2004: 318) urges us to understand that encountering the ruin is an interpretative, imaginative act: "let the ruin be the guide to the feelings within". For Ginsberg, this is a dialectical act. Caught between, on the one hand, adopting a Classical view that seeks to reconstruct from fragmentary remains a sense of the whole (ibid: 319–24), and, on the other, a Romantic view that sees the ruin not as a marker of endurance but rather as a trace of loss, the passage of time, and reminder of the irrecoverability of the past (ibid: 315–8). Holding this tension in mind is as productive for us as it is for Ginsberg, wherein ruins and paratexts are both characterised by a paradoxical invocation of both past experiences and their absence.

Anarchaeology

Our proposition in this article is that through their capacity to capture and manifest play, paratexts offer a way to reconstruct lost games and experiences. User-created paratexts, often still accessible online long after a game has disappeared, appear for us as the 'ruins' of play as it happened in the past, ready and waiting to be explored and re-animated by players and scholars alike. These paratextual ruins are traces of play that share the "fluid, processual [and] dynamic" characteristics of digital archives, where users are invited as active participants into a participatory "textual refiguring" of the past (Featherstone 2006: 596). The ruins we find within online player communities are thus, like other digital archives, "a knot of stories" that become "a live site of meaning-formation" (Carlin 2020: 179) as artefacts are consumed and accounts of play are reimagined in relation to one another. Through their vital and visceral connection to play, paratexts allow us to not only view or remember the past, but reconstitute it in new textual figurations.

To frame our reanimation of the paratextual ruins of digital games, we repurpose Thomas Apperley and Jussi Parikka's (2018) notion of 'platform anarchaeology' to outline a *paratextual* anarchaeology. Attempting to enrich platform studies by forefronting the creative and everyday practices of platform users, Apperley and Parikka gesture to the potential of paratexts to "evoke the embodied memories and experiences of users" (ibid: 354) in ways that are otherwise absent within platform studies methods. Failed, obsolete and faded platforms are the authors' explicit focus. This is because, as they explain, the examination of user-created ephemera offers a rich counterbalance to the more substantial and rigorous 'official' archives that surround the successful media platforms that typically attract scholarly attention. The platform anarchaeological approach crystallises a means by which to "speculate on alternative yet complementary trajectories for platforms" that have failed, and uses digital ephemera to piece together "speculative, alternative, minor, and even imaginary perspectives" of platform and media experiences (ibid: 360).

Repurposing these anarchaeological principles with games rather than platforms in mind builds on recent moves to turn attention to situated, living and feeling forms of archives. An anarchaeological approach to paratexts also affords a particular emphasis on speculation and multiplicity. Siegried Zielinski (2006), from whom Apperley and Parikka themselves adapt the idea of media anarchaeology, positions anarchaeology as a method firmly grounded in the imagination of alterity. In the "anarchaeological search trajectory", Zielinski notes, the historian's "place of abode is the possible" (ibid: 28), and their work "privileges a sense of the multifarious possibilities over their realities in the form of products" (ibid: 27). Eschewing "obligatory trends, master media, or imperative vanishing points", Zielinski marshals our attention instead toward the fractures and divergences that might establish a "variantology of the media" (ibid: 7). Mobility is essential to the praxis of the anarchaeologist who is tasked, in Zielinski's framing, with tracing how experiences, practices and knowledges "combine at particular moments in time, collide with each other, provoke one another" and therefore form unstable and developing processes (ibid: 258).

An anarchaeology, in the context of digital cultures, encourages us to take advantage of the fundamental "resuscitability or the 'undeadness'" of information, where ephemerality paradoxically finds permanence (Chun 2008: 171). It is by leveraging this undead persistence of otherwise transitory play experiences that paratexts – acting as traces and manifestations of past play – enable access to the "assemblage of experiences characterized both by correspondence and by difference" in sensory and affective experiences that the anarchaeological ethic seeks out (Apperley/Parikka 2018: 355). These ephemera, as we show, therefore offer us a powerful chance to move through and speculatively revive ruined, forgotten and discarded games.

There is, of course, a limit to what can be gleaned from an engagement with paratexts alone. Like ruins, paratexts provide only partial access to past experiences. Yet, as Ginsberg (2004: 158) remarks, liberated from its original utility, form and purpose by "opening its space to our presence" as visitors, "the ruin changes into itself, as we explore it". The new meaning that emerges from our exploration of a ruin "moves in directions that may have no overall unity" (ibid: 158), embracing uncertainty and imprecision. To reconstruct from paratexts then is to enlist the imagination. Herein, for Ginsberg as much as for Zielinski, lies a productive tension that avoids reconciling incongruity, instead remaining open to a plurality of possibilities and potentials.

WildStar

Developed by Carbine Studios and published by NCSoft, *WildStar* was a fantasy science-fiction Massively Multiplayer Online Role-Playing Game (MMORPG) that ran from 2014 to 2018. *WildStar* was released on 3 June 2014 under a paid-for subscription-based model, and later rereleased on 29 September 2015 as a graphically enhanced free-to-play game under the new moniker *WildStar: Reloaded* (Prescott 2015). Our analysis takes in both the subscription and free-to-play eras of the game's existence. In February 2016, the first clear signs of *WildStar*'s eventual demise emerged as NCSoft announced the closure of the game's PvP ('player versus player') servers alongside major restructuring and staff layoffs at Carbine Studios (Donnelly 2016). Two years later, on 6 September 2018, NCSoft announced the closure of Carbine Studios, with the game's servers finally being shut down on 28 November 2018 (Schreier 2018).

As a live service, *WildStar* received many patches and updates over its short lifetime. Major patches introduced, for example, new content, game modes and collectables, including what Carbine referred to as "drops" or "ultradrops," which brought about new gameplay scenarios and adventures alongside other more fundamentally game-changing content. Returning to *WildStar* through such chronicles of its development and maintenance is, for us, far less illuminating than revisiting the *affect* the game had on its players during its playable lifetime. Linear accounts chronicling the game's evolution are plentiful and can be found most succinctly detailed in the patch notes listed on community wikis (see WildStar Wiki 2017), or by revisiting the serialised press releases and news items posted on gaming news websites.

For us to carry out our paratextual anarchaeological investigation is to recognise, as Zielinski (2006: 258) notes, that such praxis best avoids chronological sequences. We resist the urge to chronologise reconstructions of past experiences with *WildStar* to instead fully embrace the manifold manifestations of play we encounter along the way. The point of this exercise, then, is not to recall what *WildStar* was, but rather to sketch out impressions of what the game felt like (or may have felt like) to play from the ruins of the player-produced paratexts that remain in our present-day.

Methodology

For the purposes of our paratextual anarchaeology of *WildStar*, we identified two research sites within which to explore the game's ruins by gathering player-created paratexts. The first of these was the *WildStar* subreddit (https://www.reddit.com/r/WildStar/), a subforum on the social platform Reddit that had 45,000 subscribers at its peak in 2014. Our second research site was the video-sharing platform YouTube (https://www.youtube.com/).

These research sites were selected because of their evident popularity among users during the time *WildStar* was accessible and playable. The two sites also offer a diversity in the types of paratext they host, and the nature of the traces of play they capture. In the *WildStar* subreddit, for example, there are to be found written reflective reports and stories, screenshots, edited images, examples of fanart and 'memes', and user-to-user discussion threads. YouTube, as is to be expected, forefronts video artefacts – whether simple recordings of play or more creative or artistic reconfigurations of game footage.

We adopted and adapted an existing game studies approach to the systematic gathering of user-created paratexts (May and Hall 2023; 2024), in which a large collection of artefacts was gathered and coded according to broad themes. Artefacts were logged in a simple research database along with key identifying information. When gathering from the r/WildStar subreddit, we excluded material related to the administration of the community and other so-called 'meta' material unrelated to experiences of the game itself. In the case of YouTube, to avoid gathering data from only the most popular videos, we utilised Bernhard Rieder's YouTube Data Tools (Rieder 2015) to organise searches by publication date, ensuring a more disinterested sample was obtained including videos with very low engagement.

A key adaptation made to May and Hall's (2023; 2024) existing approach was that a conceptual framework was not used to guide textual analysis and thematic coding. This was a deliberate response to Zielinski's concern that, often, "one is dependent upon the instruments of cultural techniques for ordering and classifying" while paradoxically seeking to "respect diversity and specialness" that such ordering obscures (2006: 27). How might we adopt the (humbling) anarchaeological orientation to "renounce power" (ibid: 27) in our approach to these paratexts? Our methodological answer was to engage with our corpus of artefacts without the narrativising influence of a conceptual framework (for example existing typologies of types of player experiences, the nature of MMO cultures, or the affective registers associated with videogame experiences) and with as little knowledge of *WildStar* itself as possible (neither of the authors has played *WildStar*). Through unfamiliarity and an unbounded analytical frame we hoped to give the paratexts themselves the space to guide us.

In practical terms, once the paratexts were captured in our database, we subjected each to textual analysis. We produced first a simple explanation of what each paratext appeared to record in terms of a play(er) experience (i.e. a descriptive exercise). Artefacts that were considered noteworthy for the richness of their textual content or particularly exemplary of patterns in the communities were then identified. This resulted in a subset of 220 paratexts, with which we undertook further textual analyses in a second phase to develop thematic readings of the player feelings or sensations evident in each (i.e. an interpretative judgement). In many ways, this second phase reflects a relatively conventional inductive qualitative process where we identified emerging themes freely through our analysis. In a third phase, we re-analysed and revised our thematic codes to make connections and consolidate similar categorisations of feelings and sensations together.

In our application of the above methods, and the identification of emergent themes, we embraced the openness and mobility encouraged by Zielinski. We heeded the anarchaeological call for the historian to "reserve the option to gallop off at a tangent, [and] to be wildly enthusiastic," seizing upon unexpected, speculative and multifarious visions of *WildStar* as they emerged in this process (Zielinski 2006: 27). The rogue archives found in the form of *WildStar*'s various paratextual remnants afford particular access, we illustrate, to the feelings and sensations that surrounded play of the game. Drawing on our corpus of paratexts, we undertook multiple reconstructions not only of what this now-defunct MMO might have been (and could have been), but also how it felt (or might have felt) to play.

As we sketch out the results of our analysis we do not identify or reference each of these paratexts directly. This is a conscious ethical decision, and reflects practice found in existing game studies paratext research (see Carter 2015; May/Hall 2023; 2024), the privacy-focused cautiousness urged in the relevant Association of Internet Researchers guidelines (2019) and principles of ethical fandom research (see Busse/Hellekson 2012; Busse 2017). While paratexts are shared in what are technically public domain venues, users themselves may have differing expectations about who their audiences are based on the cultures of these spaces. Helen Nissenbaum (2010) draws attention to a need to apply a framework that observes 'contextual integrity' when studying online communities, by attempting to respect the likely 'transmission principles' underlying users' online activities, rather than defaulting to public domain assumptions. By reporting on the artefacts we have gathered in a de-identified and often aggregated manner, we seek to strike a productive (and ethical) balance between illustrating the nature of the players' experiences with *WildStar* and exercising caution in terms of user privacy.

Our intention with a paratextual anarchaeological approach (as should be very clear by now) is not to produce a totalising or complete account of a game such as *WildStar*, but to provide insight into some of the sensations surrounding its play and players. As such, we highlight three key themes uncovered through our analysis that not only represent a significant proportion of our corpus of artefacts, but also highlight the productive tensions of this methodological approach. We illustrate in the following section these three renditions of *WildStar* and its play in the form of vignettes that capture possible textual experiences defined by *warmth*, *belonging* and *melancholy*. Our intention with these vignettes is to be deliberately imprecise: focused upon reanimating sensations and feelings, we take a coarse-grained approach to the words, images, recollections and experiences of our paratexts. Despite the systematic approach to gathering and analysing paratexts we have outlined, we deliberately eschew quantitative indications of weight or depth when journeying through our vignettes, embracing instead the fuzziness and approximation that emerges from the act of imagination (and, of course, play itself).

Reconstructions

Warmth

In our first vignette, we highlight how *WildStar*'s paratexts demonstrate players expressing *warm* feelings towards the game. We describe these feelings as derived, for those players, from an inviting sensation emanating from the game's overall stylistic presentation and appeal to nostalgia, its ability to evoke a sense of exploration and wonder, its irreverent sense of humour and joviality, and its appeal to domesticity and familiarity. Together, these form a distinctive account – or reconstruction – of *WildStar* as an MMO saturated in sensations of warmth and comfort.

Player paratexts show us that *WildStar* bore an overall lightheartedness that alleviated potential feelings of uncertainty or alienation stemming from trepidation when players were faced with the unfamiliar or unknown within the game. These disorienting aspects, we suggest, were present in both the game's premise – to explore (if not colonise) the ancient 'lost' world and alien civilisation of the Eldan – and within the necessarily complex menus and user interfaces of the game. For one poster on the r/WildStar subreddit, "the happy, cartoon-y themes in the game [...] offset just how bleak" and "absolutely terrifying" its premise was, being "so incredibly dark and frightening [hidden] behind the 'cute' characters and side-stories". For others, venturing into *WildStar*'s vibrantly coloured world, with its "cartoon-like aesthetics", "quirky adventures", sense of exploration and discovery alongside an abiding sense of "love" for how "much this game rewards us for stopping and exploring the world" were likened to feelings associated with discovering new secrets and watching Saturday morning cartoons. "*WildStar* captures," according to one commenter on the subreddit, "the spirit of old school MMOs with the great campiness that comes from these kinds of cartoons or even 80s TV". This was a "campiness" that seems to have evoked within its players a childlike wonder while at the same time instilling a disarmingly familiar sense of domesticity.

Domesticity was further embedded in many aspects of *WildStar*'s design. Not only explicitly in the game's Housing feature, which gave players an in-game site/island upon which to construct their own residence, but also implicitly, in expressions of nurture and care. Players, for example, likened managing their non-player artificially intelligent companion robots (known simply as 'bots' in-game) as the equivalent of the experience of childcare, or wrestling "two 4 year olds in a grocery store". Other players, we see, expressed desires to domesticate the cute fauna (and flora) they encountered throughout the game's virtual world (which in turn became a reality when the developers introduced a Companion Pets feature in a later update). Domesticity also manifested through one of *WildStar*'s built-in social features, where players could promote other users from the status of 'Friend' to 'Neighbor'. This granted players and their neighbours special access to visit and contribute to each others' housing allotments, complete with 'neighbourly' disputes over shared uses of resources.

Carbine Studios' choice to lean heavily into cartoonish tropes and sensibilities had other consequences for players' sensations of the gameworld as characterised by warmth. First, tonally, *WildStar*'s sense of humour and childishness befitted the archetypical characters populating its worlds. Across our gathered YouTube artefacts, for example, it is evident that quippy dialogue, adolescent humour, and humour derived from incongruity or self-depreciation (such as overt displays of authoritative incompetence and self-aware bots) all felt appropriate to players and become a core tenet undergirding the overall game's appeal. Some of the paratexts we analyse do, however, demonstrate misinterpretations of *WildStar*'s presentation as shallow, glib or worse, aimed at kids. Following Bart Simon (2017), we, along with many of *WildStar*'s players across our paratexts, consider this deliberate choice of look and feel indication that Carbine took their crafting of *unseriousness* seriously – or an obvious "labour of love", as one subreddit commenter notes.

A second influence of *WildStar*'s distinct style upon player experience emerges through its ability to conjure a sense of coherence. *WildStar*'s visual identity set it apart from its competitors, such as the gritty photorealism of Bethesda Softworks' *Elder Scrolls Online* (2014) released the same year. Yet our sense is that by adopting a cartoonish identity, *WildStar* not only managed to stand out in a very crowded marketplace but also granted Carbine liberty to have fun with its own internal logic and coherence. Nowhere is this more evident than in the ways in which the in-game movement system works, and the kinaesthetic pleasures players derive from being able to move quickly, and expressively, within the game.

Player movement in many role-playing games is often treated as mere formality, enabling players to navigate the world. Carbine Studios instead treated player movement as an essential expressive component and core experience of playing the game. Borrowing heavily from the conventions of platforming games, sprint and dash features coupled with quick and responsive controls were included. These controls meant, as we see in different playthrough videos on YouTube, that players could experience the joys of covering vast areas quickly. In one post to the r/WildStar subreddit, a player praises "the fact that you can go almost *anywhere*. [...] It feels like I'm playing a platformer", while another lauds the game's "fluid and exciting" combat system. In YouTube artefacts capturing in-game combat, we see the combination of these mechanical systems providing opportunities for players to hastily disengage or else reorient themselves during crowd control structuring combat as a fluid kinaesthetic experience.

The thrill and exhilaration associated with moving at speed were accentuated throughout the game. This is evident within YouTube videos of localised virtual events, such as one challenge players could pursue in the area of Everstar Grove titled 'Flitterfly Chaser', where collecting yellow butterflies provided the player a small speed boost. Another example – a particular characteristic of the game's Explorer character class – appears in both YouTube videos and text-posts in the game's subreddit. Distinctive flags, players in both these fora explain, are to be found throughout the gameworld and when touched grant Explorer-class players a momentary and incremental increase in movement speed. In a comic twist, when navigating the world so quickly, non-player 'bots' designed to follow the player found it hard to keep up, some even being scripted to comment pointedly (as one poster to the subreddit reports) that "a scanbot has difficulty keeping up when user is jumping all over the place."

In contrast to other MMOs like *Lineage* (NCSoft 1998) and *Guild Wars 2* (ArenaNet 2012) which strive to present a physically realistic albeit fantastical world for players to move within, Carbine's adherence to what Ward (2000) describes as "cartoonal" codes and conventions allowed them to extend the player's movement abilities beyond that which would have ordinarily been acceptable given a more realistic setting. In terms of basic movement then, it was the inclusion of a staple found in many platforming games – a double jump – that elicited the most emphatic responses among players in our gathered paratexts. Unimpressed by the pedestrian movement found in other MMOs, for example, one Reddit commentator observes that "[...] none of them felt as responsive and 'fun' to just run around in as WoW [...yet] Wildstar makes WoW feel wrong". For another YouTube commentator, "It just [struck] me that [no one] before that I know of has ever put double jump in an MMO, [even though] it's the most basic move you can have in other genres. Genius."

By embracing the freewheeling energy and expressive logic of cartoons, Carbine Studios manages to extend the verticality afforded by its double jump feature to include gravity-defying blue crystals called Loftites. One YouTube artefact we analyse demonstrates that these minerals gave players a momentary lighter-than-air sensation, and enabled them to jump even higher. Such literal levity continued over to the surfaces of the many moons that orbited the main planet, and another YouTube video illustrates the theme-park-like experience of exploring – by way of such super-powerful jumping – moons for intrepid players.

With that said, all such exaltations pale when compared to the most venerated of movement enhancements within *WildStar*: the hoverboard mount. Not only did hoverboards prove advantageous over other mounts as they enabled traversal over water, they prompted players to 'rediscover' their environments by, for example, transforming in-game terrain into the hoverboarding equivalent of skateparks. Many players fully embraced the hoverboard, with some even taking the time to post edited 'skateboarding' videos complete with stereotypical skate punk soundtracks, of them executing jumps, tricks and grinds, both on the virtual planet's surface or on the low-gravity moons orbiting it. Furthermore, many "exploited" (in the words of one user) the game's Housing feature to turn their personal residences into "radical" (in the words of another user) skateparks. Such experiments, we see in our YouTube artefacts, inspired others to combine resources in order to provide comparative boarding experiences for those playing on other servers.

Our first vignette, then, marks out a manifestation of *WildStar* that appears distinctive compared to its contemporaneous MMO counterparts. Generically orthodox mechanics, avatars, non-player characters, backstory and lore, quests, raids and opportunities for roleplay all jostle together, as they do in so many other MMOs, in our analysed artefacts. At the same time, though, *WildStar* emerges as a game fundamentally defined by expressions of jocularity, homeliness and cartoonish fun. Through the combined paratextual readings we have offered above it is apparent that for players of the game this is a rendition of *WildStar* that appears closely tied to sensations of welcoming and warmth.

Belonging

The sensations of warmth and levity we have just assembled comprise just one of the multifarious versions of *WildStar* that may have manifested through the experiences of its players. We shift our focus in this second vignette to explore another *WildStar*: one characterised by feelings of belonging, community-building, and companionship. We offer this as our second reconstruction as a means to ground our analysis in the context of the wider genre of MMOs. Drawing attention to the ways in which *WildStar* exhibits familiar dynamics and modes of engagement to its MMO-generic peers and antecedents illustrates, we hope, that this game is not an exceptional edge case. Rather, among the many renditions of *WildStar* that might be speculated about through a paratextual anarchaeology, there also lies a quotidian experience of the genre's common tendencies.

MMOs are typically designed to encourage senses of belonging and long-term player investment via socialising and meaningfully collaborating with other players (e.g. Pearce 2009; Taylor 2009). In this regard, as we sketch out below, *WildStar* is no exception. Our treatment of this particular reconstruction is, however, only a relatively brief interlude, for we are mindful of Zielinski's warning that the "great diversity" of media experiences is so often "lost because of the genealogical way of looking at things" (2006: 7). The brevity of this treatment is not intended to diminish the quality, vitality or significance of *WildStar* players' feelings of belonging as part of their play, but to acknowledge that situating the game in its generic context is precisely a genealogical exercise. The anarchaeological impetus demands we move quickly on from illustrating the game's generic underpinnings to acts of "cutting" away at this orthodoxy and investigating the "exposed surfaces" for the novel and unanticipated (ibid: 7).

There are aspects of *WildStar*'s sensations of community-building and player investment that appear, through our paratexts, dependent upon some distinct affordances of the game text itself. One feature frequently celebrated and detailed in our gathered artefacts (particularly through players' YouTube videos) is *WildStar*'s 'Path' system. In addition to choosing their faction, species, sex, and class, players were able to choose a specialist path to follow, which provided access to aspects and abilities within the game unique to each path type. For example, players choosing the 'Explorer' path gained access to otherwise inaccessible doors and platforms, enabling them to explore areas of the world other path followers could not. Collaborating with other players on different paths was encouraged through combinatory actions. For example, a Settler could repair defunct machinery, which a Scientist could then reactivate. This functionality fostered a form of collaborative social world-building unique, at the time, to *WildStar*.

Similarly, a game feature allowing players to build and decorate elaborate homes for their characters offers another insight into the game's distinctive feelings of belonging. Extending the sensations of homeliness expressed above, paratexts proudly shared by users on both YouTube and the r/WildStar subreddit showcase an array of carefully cultivated building projects. These teem with creativity and intertextual referentiality, encompassing fairytale forest settings, neon-sign-saturated retro-futurism, childlike treehouses, hoverboard skateparks, Japanese tea houses, a recreation of the *Star Wars* franchise's iconic Millennium Falcon spaceship and the 'End of Time' player base in the role-playing game *Chrono Trigger* (Square 1995). The level of deep customisation afforded in relation to *WildStar*'s player housing plots, we see, nurtures serious – and imaginative – player investment into acts of roleplay.

Other demonstrations of the experience, by some players, of *WildStar* as a text centred upon belonging are redolent of countless other encounters with MMO games. Players, for example, share fan art they have created of their in-game characters, rendered in stereotypical *anime*, 'furry', science-fiction and other genre-inflected cartoon styles. In both the /rWildStar subreddit and on YouTube we find numerous highly detailed and didactic guides for players who have reached the highest levels of character development and the most difficult late-game dungeons. One screenshot shared in the subreddit depicts 40 player avatars posing (lined up, as if in a childhood class photo) in front of an enormous – and vanquished – mechanised enemy. This post celebrates the first victory, on one of *WildStar*'s servers, in a particular endgame 'raid'. The centrality of raiding (a common mode of MMO play activity in which players must group together to fight through challenging player-versus-environment or player-versus-player encounters) is repeated across our artefacts as players reflect, in both written posts and in their commentary upon YouTube videos, upon the depth of engagement and sense of ludic investment these experiences afford.

As we noted at the outset of this second vignette, we sketch out player sensations of belonging and commonality with the wider MMO genre with some caution. There is, clearly, value in illustrating that an innocuous and archetypical encounter with the game is one of the manifold potential situated 'texts' that might emerge from play of *WildStar*. Anarchaeological caution, however, arises from the risk of becoming entangled in an account that too neatly accords, orders and classifies itself in relation to what is orthodox and expected. Consider, for example, how invocations of player investment and belonging are as central to *WildStar*'s contemporaneous marketing materials (accessible through the historical wiki websites associated with the game) as they are fundamental to MMOs and their play cultures more broadly (Pearce 2009; Taylor 2009). Resting for too long in the analytical comfort and stability of player sensations of attachment might represent an unwitting attraction to the very kind of "obligatory trend" we must avoid (Zielinski 2006: 7). As such, we move quickly now onto our third vignette, to apply Zielinski's 'cuts' to the surface of this vignette's generic familiarity in a speculative search for more of *WildStar*'s diversity and specialness.

Melancholy

Our final anarchaeological rendition of what *WildStar* might have been is contrapuntal to the two reconstructions we have already offered. In place of levity and belonging, we instead draw attention to player sensations of despair and melancholy. Consider, for example, one screenshot shared in the r/WildStar subreddit which manifests this alternative version of the game with particular potency. The image captures an in-game scene that, to most players of an MMO, would appear familiar in its frenetic and crowded visual composition. Dozens of player avatars gather near the centre of the frame, adorned with an array of the sci-fi fantasy-themed clothing, armour and weaponry that defined the game's visual style. The mass of virtual bodies is surrounded by an onslaught of extra-diegetic information: a busy server chat window, a circular mini-map, a detailed list of active quests, information about other players, user interface buttons, server statistics, and shortcuts for player actions and attacks.

What marks this screenshot as different to countless others shared by *WildStar* players is that the scene is monochromatic, rather than bursting with colour. Stark yellow text sits in the middle of the image, reading "Network Error: Connection Closed". This is because it is a screenshot taken in seconds following the 'end' of *WildStar*, when Carbine Studios disconnected the game's servers. Drained of colour, vitality and future, the image captures a rare moment of death – of a videogame itself – and is saturated instead by a melancholic affect. A closer look at the chat window reveals a long stream of players bidding their final farewells to *WildStar*, sharing messages of sorrow, gratitude and salutation to one another to the game itself.

This poignant moment in which the curtain was irrevocably brought down on *WildStar*'s five years of playable existence spurs on a number of expressions of memorialisation and melancholy across our paratexts. In another post to the r/WildStar subreddit, for example, a user shares a three-panel comic strip where dialogue from the 2018 film *Avengers: Infinity War* ("Let me guess, your home?" or "It was ... and it was beautiful") is overlayed upon images of two of *WildStar*'s characters. Another post simply titled 'WildStar is shutting down' draws 345 replies from users sharing their senses of grief and frustration, while the '"Goodbye, Wildstar" Megathread' evokes 120 comments sharing players' reflections on the highs and lows of their time spent with the game. These are feelings that are not often captured in the institutional preservation of game texts. Why, after all, labour to preserve the despair, despondency and acrimony that characterises the moment of a game's demise, when there are five years' worth of contentment, attachment, and enjoyment to turn to instead?

A paratextual anarchaeology, however, allows us to rekindle the sensations surrounding the 'end of days' of a game such as *WildStar*. It is a methodology that allows (and encourages) us to look at a moment of cataclysm such as this not as a terminal point, but as yet another opening into an alternative account of what *WildStar* 'was'. As Jenny Stümer (2023: 3) reminds us, every apocalyptic moment within which a world finds itself in peril, offers an opportunity to apprehend and "attend to the more subtle, pervasive, or slow shifts that underlie these bigger disasters". The cataclysms that mark the death and collapse of worlds function "as sites of meaning production" (Stümer 2023: 3) and, as Jill Casid (2019) argues, making sense of a crisis-stricken epoch or a way of life often first requires learning, in a sense, how to die. By placing ourselves firmly within "the scene of our undoing" and looking directly at the decay and demise presented by the 'end', we can better understand the enmeshed stories, experiences, actors, materials that comprise a world (Casid 2019: 32).

As part of an anarchaeological trajectory, we could turn our attention to the months that preceded the disconnection of *WildStar*'s servers and construct a vision of the game already steeped in its imminent demise. A series of posts made in the r/WildStar subreddit during 2013 and 2014 demonstrate the foundations of the cataclysm yet to come. Evident through the paratexts is a version of the game beset by bugs and the 'bots' used by hackers to automate cheating, and in Carbine Studios a developer either unable or unwilling to mitigate these issues. In one frustrated post, a player shares that they are engaged in a PvP match in which "I am the only real player [...] EVERY SINGLE other person on my team is a bot", eliciting more than 200 replies from other users with similar experiences to share.

A sense of helpless dismay emerges in user comments from this period of time, as glitches and errors plague player experiences, culminating for many in terminal decisions to abandon *WildStar*'s worlds entirely. In an illustrative comment posted after the game's ultimate demise, a user reflects that they were "part of the great exodus after Carbine refused to fix bugs", a situation that made ongoing participation untenable for "high end" players such as themselves. The experience of *WildStar* was, it seems, already shaped for many of its players by a melancholic recognition of the game's gradual decay.

Several paratexts capture a game in which its virtual worlds feel empty and derelict, as a consequence of the departure of disgruntled players. What one user in the subreddit describes as an "overwhelming [...] 'empty feeling world, about to die any day now' vibe" permeates other artefacts. One screenshot, for example, captures what its author alleges is the entire population of Rowsdower – one of the game's North American servers – standing together in solidarity as part of a 'strike'. Approximately thirty player-characters are arranged in the frame, protesting the developers' refusal to allow free transfers between servers or to conduct a technical act of consolidation to collapse servers together. Both of these are offered by the community as popular possible remedies to the underpopulation of the game's worlds in the wake of several 'great exoduses'. One user notes of their experience of Rowsdower that "I've only seen a single other character in a levelling zone during the last few days of playing". Where the server's capital city had been a crowded and bustling area of player activity only two months before, it now boasts "only 3-4 other people even around in the center of town." We can understand, then, that the apocalypse came to *WildStar* and defined the feeling of its play many months prior to its official shutdown on 28 November 2018.

Perhaps, though, even in reconstructing a melancholic *WildStar* by way of its apocalyptic conclusion and the ludic decay of the months immediately preceding this demise, we are still travelling in perilous proximity to the chronologies, trends and vanishing points Zielinski (2006: 7) counsels against. Zielinski, as we have noted, challenges the anarchaeologist to resist easy narratives in favour of a praxis that invites unexpected variantologies, which are best found in those particular moments in time where experiences and sensations form tensions as they "collide with each other [and] provoke one another" (ibid: 258). Searching out such tension, we drew on our corpus of paratexts to further speculate about a game that might have *always* been dying. Accordingly, an alienating and melancholic version of *WildStar* emerges through paratexts shared in the game's earlier years of service.

A handful of artefacts, for example, centre around complaints that the game's sense of 'world' and world-building is incoherent. Some users blame the way in which the game's servers manage the numbers of players visible to one another in its virtual cities (which fosters a sense of loneliness in these settlements), others the lack of narrative connectivity between different in-game areas (which render the world feeling as if it is a series of discrete, siloed locales), and yet more the challenges of communicating with other players. We might connect these despondent sensations to other accounts shared by players of their encounters with a 'toxic' culture among elite players who have reached *WildStar*'s endgame, or of the alienating experiences of new players who are confronted with dense, and tedious, tutorial levels and an overwhelming user interface. In reaching back to these earlier examples of player melancholy our point is to pull away from the simple chronology of a game that – at some point in 2018 – began to creak and crumble, and then collapse, and speculate about a *WildStar* that might never have felt 'good' to play.

Conclusion

In this article, we have proposed a method to allow defunct videogames to be revisited. Conventional digital preservation practices emphasise the role of software emulation in reconstructing *playable* videogames, but overlook the situated experiences of their players. Our intervention proposes an alternative method for more vernacular reconstructions of videogames, through recourse to what it *felt* like (or may have felt like) to play them. Our proposed 'paratextual anarchaeology' method combines Zielinski's (2006) media anarchaeology, Apperley and Parikka's (2018) subsequent platform anarchaeology, and our understanding of how paratexts operate within digital culture. In an era of user-maintained, fluid, and dynamic living archives, our method extends beyond software preservation to draw centrally upon the player-produced paratexts that persist as the 'ruins' of defunct videogames.

We aligned our proposed anarchaeological method with an understanding of paratexts as containing vital manifestations of embodied player experiences. The paratext, we hope we have demonstrated in this article, is a type of 'ruin' uniquely positioned to act as an imaginative catalyst for those revisiting play in the past. Coursing with the vibrancy of the ephemeral (as we discussed with reference to Mukherjee and others earlier), the paratext's manifestation of play is consonant with the "lively innerness", "bounding vitality" (Ginsberg 2004: 3) and "impressive immediacy" (ibid: 157) that underlies ruination. By adopting Zielinski's (2006: 28) "anarchaeological search trajectory", we sought to treat the living archives of player-produced paratexts as sites of meaning-formation, fuelled by the vitality that characterises paratextual ruins. Resisting the urge to produce master narratives, we instead strived to retain the multiplicity of perspectives afforded by our efforts to trace the feelings and sensations players expressed within the paratexts they created and shared.

We chose to experiment with this method by revisiting *WildStar*. Gathering together residual traces of play suspended in the platforms of Reddit and YouTube allowed us to arrive at three overlapping and, at times, incongruent manifestations or vignettes of what *WildStar* felt like (or may have felt like) to play during its existence as a playable text. In our first rendition of *WildStar*, defined by its jocularity, homeliness, and cartoonish fun, we identified feelings we mainly associate with *warmth*. In our second speculative reconstruction of the game, feelings of *belonging* to and participating in a virtual community were illustrated, and presented familiar sensations for those accustomed to the tendencies of the MMO genre. In contrast to this orthodoxy, in our third and final vignette, *melancholy*, we drew together the loss, despair and acrimony felt by its players at different stages of its existence and demise.

Our renditions should not be mistaken for linear and stable coverage of a game's evolution but are instead characterised by nonlinearity and fluidity. In our proposed method, we act in the spirit of Zielinski, Apperley and Parikka and do not ascribe universal meanings to the game through our speculative vignettes. Instead, given the mutable nature of living archives, we presuppose that any further scholarly visitations to *WildStar*'s ruins would introduce degrees of variance. Anarchaeology speaks to the manifold potential for multiple – at times conflicting – interpretations to coexist.

The multiplicity of remembrance offered by a paratextual anarchaeology also speaks to the multiplicity of the 'ruin' itself. As Ginsberg notes, "the ruin invents and not merely endures" (2004: 155) and encounters with these fragments of the past are characterised by a kind of freshness. This impulse for newness that lies within ruination means that as ruins reform into new "wholes" in the present – another ineluctable impulse, according to Ginsberg – these reconstructions "need have no reference to the original" experience, instead offering opportunities for a "plurality of unities" to emerge (ibid: 156). In his balancing of the so-called Romantic and Classical orientations toward ruins, Ginsberg emphasises the imaginative acts that lie at the heart of encounters with ruination. The ruin "requires discovery and stimulates exploration", instantiating a "field of happening" with the visitor (or in our case the player, or scholar) at its heart (ibid: 156). Aesthetic experiences of ruination prompt (or perhaps demand) openness, freedom and creativity from an individual seeking out and re-imagining meaning in the fragments before them (ibid: 156–7). The ruin – and the paratext – are sensed, imagined and lived anew by their visitors.

An anarchaeological orientation also helps to illustrate how the structure and logic of ruination saturates play and play cultures. The traces that games and their players leave within digital cultures are marked by the same paradoxes that characterise ruins: the simultaneity of absence and presence, destruction and generation, dullness and imagination and abandonment and inhabitation (ibid: 325–328). Every impression that playing a videogame leaves, when captured in user-created paratexts, marks both the end of an ephemeral moment of experience as well as a lively manifestation of play. Any given moment of gameplay is undead in this manner, receding almost immediately out of the player's experience as they move ever-onward with their cybernetic enactment of the game. Digital ephemera are therefore traces of the fundamental ruination that videogame play entails. Paratextual anarchaeology represents a particular orientation toward the ruins of play and games (characterised by exploration, imagination and alterity), and a means to engage this orientation as an interpretive strategy. By following an anarchaeological impetus when approaching videogames' ruins, play can be resurrected in vernacular and speculative ways appropriate to the situatedness and variety of player experiences.

References

Apperley, T./Parikka, J. (2018): "Platform Studies' Epistemic Threshold", Games and Culture, 13(4), pp. 349–369. DOI: 10.1177/1555412015616509.

Association of Internet Researchers. (2019): Internet Research: Ethical Guidelines 3.0. https://aoir.org/reports/ethics3.pdf.

Banks, J./Potts, J. (2010): "Co-creating games: a co-evolutionary analysis", New Media & Society, 12(2), pp. 253–270. DOI: 10.1177/1461444809343563.

Barker, M. (2017): "Speaking of 'paratexts': A theoretical revisitation", Journal of Fandom Studies, 5(3), pp. 235–249. DOI: 10.1386/jfs.5.3.235_1.

Belghast (2022): "Revisiting Wildstar" (29 August 2022). Retrieved from https://aggronaut.com/2022/08/29/revisiting-wildstar/.

Birke, D./Christ, B. (2013): "Paratext and digitized narrative: mapping the field", Narrative, 21(1), pp. 65–87. DOI: 10.1353/nar.2013.0003.

Busse, K. (2017): "The ethics of studying online fandom", in Click, M.A. and Scott, S. (eds.), The Routledge Companion to Media Fandom, New York: Routledge, pp. 9–17.

Busse, K./Hellekson, K. (2012): "Identity, ethics, and fan privacy", in Larsen, K. and Zubernis, L.S. (eds.), Fan Culture: Theory/Practice, Newcastle upon Tyne: Cambridge Scholars Publishing, pp. 38–56.

Carbine Studios. (2014): "WildStar", Pangyo, Seongnam, South Korea: NCSoft.

Carlin, D. (2020): "Designing for the performance of memory", in Pink, S., Ardévol, E. and Lanzeni, D. (eds.), Digital Materialities: Design and Anthropology, Abingdon: Routledge, pp. 175–193.

Carlin, D./Vaughan, L. (eds.). (2015): Performing Digital: Multiple Perspectives on a Living Archive, Farnham: Ashgate.

Carter, M. (2015): "Emitexts and paratexts: Propaganda in EVE Online", Games and Culture, 10(4), pp. 311–342. DOI: 10.1177/1555412014558089.

Casid, J.H. (2019): "Doing things with being undone", Journal of Visual Culture, 18(1), pp. 30–52. DOI: 10.1177/1470412919825817.

Chun, W.H.K. (2008): "The Enduring Ephemeral, or the Future Is a Memory", Critical Inquiry, 35(1), pp. 148–171. DOI: 10.1086/595632.

Consalvo, M. (2007): Cheating: Gaining Advantage in Video Games, Cambridge, MA and London: The MIT Press.

Consalvo, M./Begy, J. (2015): Players and Their Pets: Gaming Communities from Beta to Sunset, University of Minnesota Press.

De Kosnik, A. (2016): Rogue Archives: Digital Cultural Memory and Media Fandom, Cambridge, MA: The MIT Press.

Donnelly, J. (2016): "WildStar Devs Reportedly Lay Off 70 People", Rock, Paper, Shotgun, 14 March. https://www.rockpapershotgun.com/wildstar-carbine-studios-layoffs.

Dovey, J./Kennedy, H.W. (2006): Game Cultures: Computer Games as New Media, Maidenhead: Open University Press.

Featherstone, M. (2006): "Archive", Theory, Culture & Society, 23(2-3), pp. 591–596. DOI: 10.1177/0263276406023002106.

Genette, G. (1997): Paratexts: Thresholds of Interpretation, London: Cambridge University Press.

Giddings, S./Kennedy, H.W. (2008): "Little jesuses and fuck-off robots: on aesthetics, cybernetics, and not being very good at Lego Star Wars", in Swalwell, M. and Wilson, J.A. (eds.), The Pleasures of Computer Gaming: Essays on Cultural History, Theory and Aesthetics, Jefferson, NC: McFarland, pp. 13–32.

Ginsberg, R. (2004): The Aesthetics of Ruins. Leiden and Boston: Brill.

Giordano, F./Perron, B. (eds.). (2014): "The Archives: Post-cinema and Video Game Between Memory and the Image of the Present", Milan: Mimesis International.

Gray, J. (2010): Show Sold Separately: Promos, Spoilers, and Other Media Paratexts, New York: New York University Press.

Guay-Bélanger, D./Deslongchamps-Gagnon, M./Lavigne, F./Perron, B. (2023): "Game(play) Archives: Quebec Video Games as Case Study", G|A|M|E Games as Art, Media, Entertainment, 1(10).

Guins, R. (2014): Game After: A Cultural Study of Video Game Afterlife, Cambridge, MA: The MIT Press.

Guttenbrunner, M./Becker, C./Rauber, A. (2010): "Keeping the Game Alive: Evaluating Strategies for the Preservation of Console Video Games", International Journal of Digital Curation, 5(1), pp. 64–90. DOI: 10.2218/ijdc.v5i1.144.

Harkin, S. (2024): "Archival Challenges for Inclusive Games History", Proceedings of DiGRA Australia 2024, presented at DiGRA Australia 2024, Melbourne.

Janik, J. (2018): "The Ruins of Meridian 59: Abandoned MMOs as Poor Objects". DiGRA 2018, Turin, Italy, July 2018. Digital Games Research Association.

Kirkpatrick, G. (2012): "Constitutive Tensions of Gaming's Field: UK gaming magazines and the formation of gaming culture 1981-1995", Game Studies, 12(1). http://gamestudies.org/1201/articles/kirkpatrick.

Kocurek, C.A. (2015): Coin-Operated Americans: Rebooting Boyhood at the Video Game Arcade, Minneapolis and London: University of Minnesota Press.

Lowood, H. (2002): "Shall we Play a Game: Thoughts on the Computer Game Archive of the Future", Bits of Culture Conference, Stanford University, pp. 1–19.

Lowood, H. (2006): "High-performance play: The making of machinima", Journal of Media Practice, 7(1), pp. 25–42. DOI: 10.1386/jmpr.7.1.25/1.

Lowood, H. (2011): "Perfect Capture: Three Takes on Replay, Machinima and the History of Virtual Worlds", Journal of Visual Culture, 10(1), pp. 113–124. DOI: 10.1177/1470412910391578.

Maclean, M. (1991): "Pretexts and Paratexts: The Art of the Peripheral", New Literary History, Johns Hopkins University Press, 22(2), pp. 273–279. DOI: 10.2307/469038.

Manning, J. (2017): "Unusable Archives: Everyday Play and the Everyplay Archives", in Swalwell, M., Stuckey, H. and Ndalianis, A. (eds.), Fans and Videogames: Histories, Fandoms and Archives, New York and London: Routledge, pp. 197–213.

Manoff, M. (2004): "Theories of the Archive from Across the Disciplines", Portal: Libraries and the Academy, 4(1), pp. 9–25. DOI: 10.1353/pla.2004.0015.

May, L. (2021): Digital Zombies, Undead Stories: Narrative Emergence and Videogames, New York: Bloomsbury Academic.

May, L./Hall, B. (2023): "From aesthetics to asymmetry: Contradictions of ecological play in Cities: Skylines", Games and Culture. DOI: 10.1177/15554120231219729.

May, L./Hall, B. (2024): "Thinking ecologically with Battlefield 2042", Game Studies, 24(1). https://gamestudies.org/2401/articles/mayhall.

Morris, S. (2003): "WADs, Bots and Mods: Multiplayer FPS Games as Co-creative Media", Level Up Digital Games Research Conference.

Mukherjee, S. (2015): Video Games and Storytelling: Reading Games and Playing Books, Basingstoke, Hampshire and New York, NY: Palgrave Macmillan.

Murphy, D.L. (2015): "Documenting Pocket Universes: New Approaches to Preserving Online Games", Preservation, Digital Technology & Culture, De Gruyter Saur, 44(4), pp. 179–185. DOI: 10.1515/pdtc-2015-0021.

Newman, J. (2011): "(Not) Playing Games: Player-Produced Walkthroughs as Archival Documents of Digital Gameplay", International Journal of Digital Curation, 6(2), pp. 109–127. http://www.ijdc.net/index.php/ijdc/article/view/186.

Newman, J. (2012): Best Before: Videogames, Supersession and Obsolescence, London: Routledge.

Nissenbaum, H. (2010): Privacy in Context: Technology, Policy, and the Integrity of Social Life, Stanford, CA: Stanford University Press.

Nooney, L./Driscoll, K./Allen, K. (2020): "From Programming to Products: Softalk Magazine and the Rise of the Personal Computer User", Information & Culture, 55(2), pp. 105–129. DOI: 10.7560/IC55201.

Nylund, N. (2015): "Walkthrough and let's play: evaluating preservation methods for digital games", Proceedings of the 19th International Academic Mindtrek Conference, New York, NY: Association for Computing Machinery, pp. 55–62. DOI: 10.1145/2818187.2818283.

Pearce, C. (2009): Communities of Play: Emergent Cultures in Multiplayer Games and Virtual Worlds. Cambridge, MA and London: The MIT Press.

Pinchbeck, D./Anderson, D./Delve, J./Otemu, G. (2009): "Emulation as a strategy for the preservation of games: the KEEP project", DiGRA '09 – Proceedings of the 2009 DiGRA International Conference: Breaking New Ground: Innovation in Games, Play, Practice and Theory, pp. 1–5.

Potts, J./Hartley, J./Banks, J./Burgess, J./Cobcroft, R./Cunningham, S./Montgomery, L. (2008): "Consumer co-creation and situated creativity.", Industry and Innovation, 15(5), pp. 459–74. DOI: http://doi.org/10.1080/13662710802373783.

Prescott, S. (2015): "WildStar is getting graphical enhancements when it goes free-to-play", PC Gamer, 25 September. https://www.pcgamer.com/wildstar-is-getting-graphical-enhancements-when-it-goes-free-to-play/.

Rieder, B. (2015): "YouTube Data Tools". https://ytdt.digitalmethods.net

Schreier, J. (2018): "WildStar Developer Carbine Studios Shuts Down", Kotaku, 6 September. https://kotaku.com/wildstar-developer-carbine-studios-shuts-down-1828862729.

Scully-Blaker, R. (2023): "Reframing the Backlog: Radical Slowness and Patient Gaming", in De Beke, L.O., Raessens, J., Werning, S. and Farca, G. (eds.), Ecogames, Amsterdam: Amsterdam University Press. DOI: 10.5117/9789463721196_ch24.

Simon, B. (2017): "Unserious", Games and Culture, 12(6), pp. 605–618. DOI: 10.1177/1555412016666366.

Stümer, J. (2023): "Introduction: understanding apocalyptic transformation", in Stümer, J. and Dunn, M. (eds.), Worlds Ending. Ending Worlds, Berlin: De Gruyter, pp. 1–18. DOI: 10.1515/9783110787009-002.

Švelch, J. (2020): "Paratextuality in Game Studies: A Theoretical Review and Citation Analysis", Game Studies, 20(2). http://gamestudies.org/2002/articles/jan_svelch.

Swade, D. (2002): "Collecting Software: Preserving Information in an Object-Centred Culture", in Hashagen, U., Keil-Slawik, R. and Norberg, ArthurL. (eds.), History of Computing: Software Issues SE – 19, Berlin: Springer, pp. 227–235. DOI: 10.1007/978-3-662-04954-9_19.

Taylor, T.L. (2009): Play Between Worlds: Exploring Online Game Culture, Cambridge, MA: The MIT Press.

Therrien, C. (2015): "Inspecting Video Game Historiography Through Critical Lens: Etymology of the First-Person Shooter Genre", Game Studies, 15(2). http://gamestudies.org/1502/articles/therrien.

Thibodeau, K. (2012): "Wrestling with shape-shifters: perspectives on preserving memory in the digital age", Proceedings of 'The Memory of the World in the Digital Age: Digitization and Digital Preservation, pp. 15–23.

Walker, A. (2019): "The History of Games Could Be a History of What Play Felt Like", ROMchip, 1(1). https://romchip.org/index.php/romchip-journal/article/view/78.

Ward, P, (2000): "Defining 'Animation': The Animated Film and the Emergence of the Film Bill", Scope: An Online Journal of Film Studies, pp. 1–11.

Webber, N. (2023): "The past as (para)text – relating histories of game experience to games as texts", in Seiwald, R. and Vollans, E. (eds.), (Not) in the Game: History, Paratexts, and Games, Berlin: De Gruyter, pp. 81–98.

WildStar Wiki. (2017): "Patch notes", WildStar Wiki. https://wildstar.fandom.com/wiki/Patch_notes.

Winget, M.A. (2011): "Videogame preservation and massively multiplayer online role-playing games: A review of the literature", Journal of the American Society for Information Science and Technology, 62(10), pp. 1869–1883. DOI: 10.1002/asi.21530.

Zielinski, S. (2006): Deep Time of the Media: Toward an Archaeology of Hearing and Seeing by Technical Means, Cambridge, MA: The MIT Press.

Post-apocalyptic ruins in digital games as indexical storytelling devices

Romi Sofia Abatangelo

Abstract

This paper investigates how ruins are used in digital games as indexical storytelling devices, referring to a form of storytelling in which traces are used to point to the events that have generated them, effectively turning the player into a detective whose goal is to reconstruct what happened. After discussing in the first part of the paper the connections between ruins and gameworlds, the narrative potential of ruins, and the concept of the post-apocalyptic, the second part of the paper will analyse as case studies the games Horizon Zero Dawn *(Guerrilla Games 2017) and* Disco Elysium *(ZA/UM 2019), in which the use of post-apocalyptic ruins as indexical storytelling devices holds a prominent role in the construction of the game narrative.*

Keywords

environmental storytelling, indexical storytelling, ruins, post-apocalyptic

Digital ruins as storytelling devices

> "The beauty of ruins is not given. It is found."
>
> ROBERT GINSBERG – THE AESTHETICS OF RUINS

In his inquiry about the relationship between ruins and digital games, Vella (2010) makes a broad distinction in regards to how ruins are integrated into video game spaces. On one hand, ruins can be used as a trope to create an exotic atmosphere, to evoke the incomparable beauty of a lost Golden Age, or to remind us of our own mortality and the impermanence of our civilisation. On the other hand, the integration of ruins in digital games can go beyond the thematic aspect and fully reproduce the *ruin-situation*, described as "the interpretative framework that the viewer interposes between herself and the ruin, in order to decipher the ruin into a comprehensible entity" (Vella 2010: 44). The ruin-situation is then mapped-out as an intersection of conflicting forces: on one hand the relationship between the ruin-site and the viewer, one possessing the power for signification, the other exploring and extrapolating; on the other hand the conflict between two different interpretations of ruins as vestiges of a lost original totality, or as survivors of the events that led to their ruination. The ruin-situation thus describes the experience of ruins firstly in regards to the mode of fruition, and secondly as pointing out to multiple interpretations.

In regards to the first aspect, Vella discusses the similarity between the mode of fruition of the ruin-site and the game space, referring in particular to adventure games understood as "featuring, as a fundamental constitutive element, the negotiation and traversal of the gameworld by an avatar representing the player's embodiment within that world" (ibid: 54). Drawing from Ginsberg's (2004) description of the aesthetic experience of ruins – incorporating the participant, requiring discovery and stimulating exploration – Vella constructs a parallel with the player experience of gameworlds, described as an "unordered topography that the player is required to map out and bring into some form of coherence through the interpretative work of plotting lines of movement and action across it" (Vella 2010: 60). In order to construct a framework that could describe both ruins and games, Vella proposes the concept of *explorable spaces*, "disordered spaces, embedded with fragments or suggestions of forms to be pieced together" (ibid: 61). Exploration is thus considered not only in its geographical sense but also as an intellectual inquiry, an activity of interpretation and construction of meaning. In regards then to the narrative potential of ruins, Vella highlights the similarity between the semiotic potential of the ruin-site – bearing traces of both its original totality and the process of ruination – and the *embedded narrative structure*, a particular narrative structure often employed in digital games. First discussed by Jenkins (2004), the embedded narrative structure consists in the creation of a predetermined narrative conveyed through fragments and clues that are embedded into the game space, waiting for the player to discover and interpret them as they explore. As Vella points out, this structure "frames the gameworld as a repository of traces of the past, a tissue of its own history […]. This past-orientedness – bearing the implication that the gameworld can only be deciphered by reference to past events – immediately suggests the ruin as the natural home of the embedded narrative." (Vella 2010: 72). Ruins are the remains of a past that is no longer there, the survivors of the inevitable process of decay, and thus they can act as a bridge between the past and the present, between the story that is embedded and the here-and-now of the viewers.

Jenkins' conceptualization of embedded narrative has since been picked up by other scholars, with the aim to expand and further understand it. Domsch (2019) identifies different ways in which narrative information can be embedded in the game space, differentiating between embedded narratives, "understood [as] all kinds of explicit narrative content that a player encounters while navigating the world of a videogame" (Domsch 2019: 113) and visual clues, "defined as any kind of visually detectable signs within a videogame's navigable space that has narrative potential" (ibid: 108). Fernández-Vara (2011) instead investigates the semiotic relationship between visual clues and their meaning. Referring to Charles Peirce's classification of signs, the scholar proposes the concept of *indexical storytelling* to describe a form of narrative that is mostly conveyed by visual clues:

"An index has a relation to the event, often is the consequence of it, which points to something that happened or is going on, inviting the player to reconstruct what happened. The player has to connect the relationship between the sign and the event that it points to." (Fernández-Vara 2011: 4)

The concept of indexical storytelling can then be used to discuss the ruin-situation as described by Vella (2010). As meaning in indexical signs emerges through their power to point towards an event, past or ongoing, ruins can be considered as indexes, pointing towards the event leading to their ruination. However, considering indexes in a more general sense as signs in which "the idea is physically connected with the sign" (Fernández-Vara 2011: 4) it can be said that ruins possess an indexical relationship not only with the agent of ruins, but also with the original state of the ruin-site before the process of decay. As street signs pointing towards a destination, the ruined remains point towards their past glory and wholeness, however constructing a relationship that is not spatial but temporal. Ruins then generate a peculiar form of indexical storytelling, simultaneously pointing towards two different areas of signification. Along with the non-linearity and fragmentation generated by a fruition based on discovery and exploration, this tension between multiple interpretations further enhances the sense of mystery and openness that is characteristic of environmental storytelling.

Post-apocalyptic ruins

> "Some things can't be fixed. Some things lost are lost forever. No matter how hard we fight, how much it hurts, or how much our hearts yearn to put them back together."
>
> J. D. Payne & Patrick McKay – The Lord of the Rings: The Rings of Power

The Book of Revelation, also known as the Apocalypse of John, is the last book of the New Testament. Written around 90 A.D. by John of Patmos, a Jewish prophet escaping the war ravaging Judea at the time, the book contains a detailed prophecy of the end of human civilisation, as God announces a war against evil that will destroy the universe in the process. However, the prophecy also foretells that God's ultimate victory will lead to a Golden Age, in which the righteous among men will live in a New Jerusalem descended from the sky (Pagels 2012). The word apocalypse – from the Greek *apokalupsis*, meaning revelation or unveiling of the true order (Heffernan 2008) – then started to be used at first to indicate the events leading to end of the world, and then by extension to indicate "a disaster resulting in drastic, irreversible damage to human society or the environment, especially on a global scale" ("Oxford English Dictionary: Apocalypse" n.d.). Heffernan (2008) discusses how an apocalyptic worldview – in which "History, the Nation, and Man [...] continue to be secured by the spirit of the Christian apocalypse, a narrative that posits an origin and moves definitively, through a series of coherent and concordant events, towards an end that will make sense of all that has come before it" (Heffernan 2008: 4) – is challenged by many twentieth-century narratives, as the devastation of the First World War and the horrors of the Second World War contributed to the shift from an apocalyptic towards a post-apocalyptic worldview in modernist and post-modernist literature. While the destruction brought by the apocalypse is always followed by a new and better order of things, post-apocalyptic narratives offer no salvation, only destruction. Survivors of the post-apocalypse are left to deal with the consequences of the cataclysm, living within the ruins of the world as it was. Consequently, post apocalyptic narratives – especially in visual media such as film (see Määttä 2019) and video games (see Fraser 2019) – heavily rely on the presence of ruins to evoke the catastrophe that lead to the current state of the world, as well as the state of future decay of our present civilisation. In that regard, while the post-apocalyptic as a literary genre is generally set in a dystopian future and thus deals with the ruins of our present, in this paper I will use the term post-apocalyptic to indicate any fictional world in which a catastrophic event has taken place in the past, radically transforming the state of the world so that a before and after can be clearly identified: the world after is on one hand bearing the scars of the catastrophic event and on the other hand disseminated of fragments of the world before.

In the previous section I have mentioned how ruins bear traces of the agent of their ruination. In Vella's ruin-situation model, the term *ruin-agent* describes a broad range of phenomenon: "it can refer to a physical enemy, such as the Goths sacking Rome; more abstractly, it can refer simply to natural processes of decay, or even to time itself – in short, anything that arrays itself in opposition to the ordered sphere set up by the original construction, and that contributed to its ruination" (Vella 2010: 45). Ginsberg (2004) broadly distinguishes between decay – namely the form of damage that comes with ageing – and sudden destruction by either natural catastrophe or human deed. Decay is framed as an inevitable process – the slow corrosion of material and invasion by natural life – that civilisation needs to contrast with constant maintenance. Catastrophic events, on the other hand, violently and suddenly damage a site, transfiguring it in a short lapse of time. These two modes of ruination produce very different emotional effects: while the presence of natural life growing on the remains of the past can generate a sense of peace and acceptance, the sudden ruination produced by a catastrophic event leaves a deep scar in the site and creates instead a sense of dread and commiseration.

In the following section of this paper, I will analyse two case studies showcasing the different impact between a game world that represents the ruins of a past civilisation that have been incorporated by nature and one that represents the ruins of a bombarded city that has been left in a permanent state of disarray. Set into a post-apocalyptic version of the Southwest of the United States, returned to nature centuries after the fall of Western civilisation, *Horizon Zero-Dawn* (Guerrilla Games 2017) employs environmental storytelling techniques to construct a dialogic relationship between the sense of loss of our heritage, the innocent beauty of nature and the impending danger that hides behind the surface. Set into the postcolonial district of Martinaise, *Disco Elysium* (ZA/UM 2019) constructs instead a dialogue between the ruination of war and the resilience of the inhabitants, repurposing the damaged urban fabric in spite of the lack of investment and aid from the occupying forces.

Horizon Zero Dawn: the ruins of the ecological apocalypse

> "Look around you, at today's world. Your house, your city. The surrounding land, the pavement underneath, and the soil hidden below that. Leave it all in place, but extract the human beings. Wipe us out, and see what's left."
>
> ALAN WEISMAN, THE WORLD WITHOUT US

In the global bestseller *The World Without Us,* the award-winning journalist Alan Weisman explores the impact of the human population on our planet, imagining how the world would react to the sudden disappearance of humanity. How long will it take for nature to reclaim its space? How will human architectural and engineering works degrade as power and maintenance are no longer supplied? Which species of plants will grow on the abandoned concrete? How long will the ruins of our monuments and statues take to fully degrade? Will the world ever be able to return to its pre-human state, or did our civilisation alter the planet so irreversibly that our traces could never fully be erased? The game world of the action-adventure role-playing game *Horizon Zero Dawn* (2017) creates its own version of a world without us. Set in the 31st century ("Horizon Wiki: Timeline" n.d.), the game portrays a post-post-apocalyptic civilisation (Condis 2020) born over the ruins of the Old World, namely the neoliberalist western civilisation of the 21st Century that was completely destroyed by an apocalyptic event of gigantic proportions. A rogue army of military robots started self-replicating using biomass as fuel, inevitably causing the destruction of all life forms on Earth before scientists could regain control of them. An artificial intelligence named GAIA was created before the apocalypse, with the task of deactivating the military robots, then guiding the terraforming system to restore life on the planet and raising a new generation of humans. These new humans were then released into a world in which nature has grown over the ruins of the Old World.

While *Horizon Zero Dawn* is set into a fictional post-post-apocalyptic future, is it possible to identify real-world scenarios that inspired the aesthetic of its game world, in particular the relationship between ruins and nature. In that regard, Ginsberg (2004) discusses the complex relationship between architectural remains and spontaneous vegetation in the ruins, highlighting the negotiation between natural growth and human management of the ruin-site. But what happens to ruins in the absence of human intervention, in a *world without us*? The exclusion zones of Chernobyl and Fukushima provide a small-scale example of the activity of nature in the absence of humanity. Both exclusion zones are the result of accidents in nuclear power plants. The Chernobyl nuclear accident happened in 1986 in Ukraine, when an improperly conducted test caused an explosion in one of the reactors. The Fukushima accident happened in 2011, after a tsunami wave went beyond the 10m wall protecting the nuclear power plant, flooding the site and damaging the reactor's cooling system. While the accidents had widely different environmental impacts, in both cases exclusion zones were created in the areas where the radiation levels were considered unsafe for the human population (Steinhauser et al. 2014). The eerie landscape of Chernobyl's 30 km radius exclusion zone still holds a prominent role in the collective imagery, becoming the the setting of many of many post-apocalyptic narratives – such as the video game adaptations of Tarkovsky's movie *Stalker* (Stone 2013) – and the subject of various art projects – such as the interactive installation *Post-Apocalypsis* (Jelewska/Krawczak 2018) and the photoshoot *Legacy: Inside the Chernobyl Exclusion Zone* (Darwell 1998). It is also a dark-tourism destination, with illegal tours being organised even prior to the Ukrainian government authorisation for official tours in 2011 (Stone 2013). The site of Pripyat, in particular, is an interesting case study to discuss the relationship between abandoned architecture and nature. Also called Atomgrad (the city of the atom), the town of Pripyat was built in the 1970s to house the Chernobyl nuclear plant workers, and was subsequently evacuated after the accident. Its brutalist architecture has been described as an "example of blocky architectural Brezhnev baroque [turned by the nuclear accident] from an icon of modern Soviet planning and technology into an icon of Soviet political ineptness, bureaucratic incompetence, and technological calamity" (Stone 2013: 82-83). Tourists visiting Pripyat witness the ghost of humanity, encountering the traces left behind by the former inhabitants (Rush-Cooper 2020), but they are also met with a new growing presence. Undisturbed by human activities, plants are sprouting over the cement of pavements and buildings alike, slowly turning the brutalist architecture into a forest. In the absence of culture, nature has claimed the ghost town for itself.

"In Pripyat, an unlovely cluster of concrete 1970s high-rises, returning poplars, purple asters, and lilacs have split the pavement and invaded buildings. Unused asphalt streets sport a coat of moss. In surrounding villages, vacant except for a few aged peasants permitted to live out their shortened days here, stucco peels from brick houses engulfed by untrimmed shrubbery. Cottages of hewn timbers have lost roof tiles to tangles of wild grapevines and even birch saplings." (Weisman 2007: 216)

The exclusion zone of Fukushima, albeit smaller and more recent, shows how fast nature reclaims the space that humanity has carved out for itself. The short film *Contaminated Home* (Fisher/el Sani 2021) follows a Japanese couple – evacuated to Kyoto after the accident – periodically returning to their house inside the evacuation zone with a photo camera and a Geiger counter to document the level of radiation and the transformation of their home. Their photos, as well as the visits with the documentarists, show lush vegetation invading what once was a carefully landscaped Japanese garden now claimed back by the forest. Despite the flourishing nature, however, the ghost of the nuclear accidents is still haunting the verdant forests. Invisible but lethal, radioactive pollution renders nuclear exclusion zones unsafe for humans. Wildlife has returned to Chernobyl not long after the accident and – in the absence of the human population – have not only survived but thrived. However, while radioactivity did not destroy life, it left the animal population damaged, corrupting their genomes and impacting their survival abilities (see Weisman 2007). Invisible and yet lethal, radioactivity haunts the ground in Chernobyl and Fukushima, a hidden threat that will likely survive any warning signs and interdiction, poisoning animals and people that will reclaim the land as their own in the future.

The game world of *Horizon Zero Dawn* presents a similar relationship between ruins, nature and the traces of the catastrophe. In the New World – terraformed by GAIA – the ruins of the Old World are immersed into a natural landscape of breathtaking beauty, enhanced by the photorealistic graphics and the immersive sound design:

"Horizon basks its players in the experience of sublime landscapes of mountain rises, canyons, hills, rivers, meadows of high-waving grass used for stealth, the visceral rendering of rain or the realistically audible sound of crunching snow underfoot." (Falkenhayner 2021: 7)

However, like the Exclusion zones of Chernobyl and Fukushima, this post-apocalyptic paradise is haunted by an invisible threat. Not poisonous radiations lingering on the ground, but indestructible military robots buried under the terraformed surface, ready to strike again if given the right order. While the new human inhabitants of the world do not possess the necessary technology to take control of the robots, the New World is haunted by another remnant of the Old World. The same artificial intelligence that deactivated the rogue military robots and controls the terraforming system that restored the planet to its natural state – a conglomerate of AIs controlled by GAIA – has the potential to destroy everything again. At the start of the game, the idyllic world – in which GAIA's animal-like machines live peacefully among humans and wildlife, performing their duty of maintaining the terraforming system – has started to reveal this hidden threat. Machines have turned hostile, attacking people and forcing them outside their pastures and in return being hunted for parts by human tribes. During the game this threat will progressively reveal as the player accompanies the protagonist Aloy – a young woman generated by GAIA using the DNA of her creator, the scientist Elizabeth Sobek – in her quest to save the world.

Players then are able to explore the game world as Aloy's journey unfolds, encountering the remains of the Old World as well as the new civilisation that has developed in the New World. While most of the human population has limited to no access to the Old World's technology, at the very beginning of the game Aloy finds a *focus*, a device that allows her to access a layer of augmented reality, providing access to embedded narrative content (Domsch 2019) in the form of textual and audio logs and holographic projections. This AR layer is new and mysterious to Aloy, but familiar to players, who bring their own perspectives to the fictional gamespace (Aghoro 2022). Attached to the unique point of view of the player-character, players are then able to navigate between three different layers of temporality: the New World in which the events of the game happen, the futuristic technology of the Old World found in bunkers and cauldrons, and the remains of the non-fictional western civilisation as ruins scattered throughout the game world. Between uncovering the secrets of a futuristic past and inhabiting a tribal future, the players encounter the ruins of their everyday life. Traffic lights in the middle of the woods, collapsed highways, rusted remains of cars, helicopters, military tanks. The metallic frames of empty billboards, a rusted swing among tall blades of grass. The poles of wind turbines, their blades bent beyond repair. Gutted skyscrapers with metal frames and cement cores exposed, inhabited by moss and vines, surrounded by poplars and birches. Iconic landmarks of the Southwestern United States are turned into ruins ("Horizon Wiki: List of Real Locations in Horizon Zero Dawn" n.d.), their shapes barely recognisable. While the futuristic ruins inside bunkers and cauldrons still construct a relationship with nature – with healing mushrooms and herbs growing inside them as well as the omnipresent vines – the ruins of our civilisation are the ones in which this relationship is more prominent. Scattered among the gameworld, with no boundaries between the surrounding forests, sands or snowfields, these ruins bear a strong similarity to pictures from Chernobyl's exclusion zone. Darwell's (1998) series of photographs portray familiar objects in an uncanny state of abandonment and decay: a rusted fire truck in the middle of a field of yellow wild flowers, a flock of grey helicopters left on the ground, the metal frame of a ferris wheel, a slide inside an overgrown playground. The architecture of Chernobyl – brutalist buildings and wooden cabins alike – is better preserved than the ruins in *Horizon Zero Dawn*, but is in a similar relationship with nature, with wild plants and trees reclaiming their space. Inside the abandoned homes, the scattered belongings of the evacuees tell an unordered story, of life and death, of fear and abandonment (Rush-Cooper, 2020). Environmental storytelling through everyday objects – common in the narrative design of digital games, especially walking simulators such as *Gone Home* (Şengün 2017) – is however not employed in *Horizon Zero Dawn*. Objects of the old world are instead used as resources. Too distant to the inhabitants of the new world to retain their meaning and function, the objects that Aloy finds among ruins – old bracelets, watches, mugs – are either sold as trinkets or gathered as collectibles, as the new inhabitants pose questions about their long lost functionalities.

In conclusion, the ruins of western civilisation in *Horizon Zero Dawn* fulfil an important function in conveying and enriching the message of the game despite having little to no relationship with the main storyline. As the narrative provides a critique of eco-modernist solutions to the climate crisis (Condis 2020), the ruins of our civilisation serve as a *memento mori*, a reminder of the fragility of modern civilisation against the danger of consumerism and capitalistic exploitation of technology.

Disco Elysium: living among the ruins of war

> "The cataclysm has happened, we are among the ruins […]. We've got to live, no matter how many skies have fallen."
>
> D. H. Lawrence, Lady Chatterley's Lover

Since the beginning of time, war has always brought destruction to the architectural landscape. History is filled with tales of revolts, sacks and sieges in which the fury of battle has brought destruction to cities and their population. However, from the 18th century onwards both the destructive potential of military technologies and the civilian involvement in wars have increased exponentially, creating ruinscapes unlike anything seen before. Artillery fire and aerial bombing can tear buildings apart and set them on fire. An entire city can be razed in a matter of hours, leaving behind piles of rubble and skeletons of cement. Post-war reconstruction is a long and complex process, and cities can bear the traces of war for decades. The adventure role-playing video game *Disco Elysium* – mixing detective stories, occult and science fiction with historical references to the colonisation of America, the French and Russian revolutions, and the military operations of NATO in Eastern Europe and in the Middle East – conveys its narrative through environmental storytelling, as the ruination of the gameworld reveals its layered history. The game is set in the fictional district of Martinaise, that has been bombed by a foreign army – the Coalition of Nations – in order to repress a communist revolution and enforce economic liberalism, and is under the control of the Coalition almost 50 years later. The district, still bearing damage from the bombings, has been left to its own by both the occupants and the local police force – the Revachol Citizen Militia (RCM) – and is self-governed by the Union of Dockworkers of the Industrial Harbour. The player character is an amnesiac RCM detective called Harrier "Harry" Du Bois, who is tasked to solve a murder case with his partner Kim Kitsuragi. During this murder investigation, the physical exploration of Martinaise looking for clues will lead to the discovery of the past history of the world and its connection with the current state of affairs.

The post-colonial status of Martinaise is visually conveyed through the gameworld by the damages caused by the aerial raid that started the military occupation, evoking the powerful imagery of the urban destruction caused by armed conflicts since the 20th century. With the development of aerial warfare, the devastation of war was brought into the heart of the conflicting nations: strategic bombing was employed since the First World War to raid enemy factories and cities, in order to decrease their productive and logistic capacity and to impact the morale of the civilians (Barros 2009). Despite numerous efforts from international bodies to regulate aerial warfare, placing bans on bombing non-military targets and undefended cities (Jensen/Eaton 2018), terror bombing was almost a natural consequence of the radicalisation of ideologies that led to the dehumanisation of the enemy populations (Centeno/Enriquez 2016). The first use of terror bombing against a European city is the infamous bombing of Guernica in 1937 during the Spanish Civil War, which was carried out in such a way as to maximise civilian casualties and destruction of architectural and cultural heritage (Jensen/Eaton 2018; D'Orsi 2007). During the Second World War, strategic area bombing – namely "the strategy of treating whole cities and their civilian populations as targets for attack by high explosive and incendiary bombs" (Grayling 2006: n.p.) – was used extensively by the German and British air forces, caught in a spiral of revenge raids in which the focus shifted from military targets to the civilian population (Centeno/Enriquez 2016; Grayling 2006). Among many others, the London Blitz, the Coventry raid, the destruction of Dresden, the area bombing campaign of Berlin and the Allied bombings on Italy left a mark on the collective memory as well as a trail of ruins, targeting not only civilians but also cultural heritage in an attempt to destroy the enemy's way of life as well as its military capabilities (Grayling 2006). After the Second World War, strategic area bombing became an integral and fundamental part of warfare in local conflict up to present times. Thanks to the media, even local conflicts assume a global dimension as aerial raids are often broadcast live, while pictures of the ruined war zones are featured on the front pages of newspapers: those among us who are old enough may remember the images of the NATO bombing of Belgrade (1999) during the Yugoslavian wars, the Coalition aerial raids on Baghdad (2003) during the Iraq war, or the destruction of Aleppo (2012-2016). As I write this article, the news is filled with images of bombings from the Russo-Ukrainian War and pictures of the Gaza Strip in ruins. Even the ones who had the fortune to not live through armed conflicts are thus familiar with the imagery of ruin and devastation of the war, with the struggle of people who have to live in what's left of their homes torn apart by the bombing or in refugee camps. But what happens after? What happens after the bombings, after the war, after the media attention is directed elsewhere? The process of reconstruction is neither fast nor easy, especially considering the extensive and widespread damage caused by area bombings. People have to keep living among the ruins for years, finding shelter inside the least damaged apartment buildings or seeking refuge in shantytowns.

Roberto Rossellini's *Germania Anno Zero (Germany Year Zero,* 1948) – shot in the aftermath of the Second World War – provides a powerful testimony not only of the damage to the urban landscape but also of the social impact of the war. Filmed in Berlin in 1947, *Germania Anno Zero* revolves around the story of a young child, Edmund, following his lack of moral guidance and his struggle for survival in the capital city of the country that lost the war (Rondolino 1989). The Allied bombings turned Berlin into a ruinscape of gutted buildings – their brick-made skeletons exposed, the few surviving palaces as memories of a lost splendour – the architectural ruins mirroring the ruination of Edmund's childhood. As the boy ultimately commits suicide by jumping from the roof of his house, he "become[s] part of the rubble, both symbolic and literal, that has been his universe: his body lies, shattered, at the base of a ruined building." (Craig, 2010: 66). As area bombing as a war strategy constructs a relation of power between the destructive power of aerial warfare and the powerlessness of civilians (D'Orsi 2007), a similar relation of power emerges in the aftermath: the shattered lives of the survivors, who lost not only their loved ones but also their previous way of living, are subjected to the whims of the occupying forces, whose favours become necessary for survival. In Rossellini's movie, the ruins of the city mirror the moral and material ruins of their inhabitants, from the young women forced to prostitute themselves to the occupiers, to the hustling boys dealing with the deaths and the struggles of their family members. The dynamic of the occupation is similarly represented in *Dysco Elysium*: as the post-colonial Martinaise has been left to its own fate, the lives of the people reflect the ruination of the district. Despite the ultraliberal regime enforced by the occupiers, businesses struggle to stay open in Martinaise, while the ruined district becomes a home for the outcasts of society: the poor, the addicts, the outcasts. Like in Rossellini's *Germania Anno Zero*, children in *Disco Elysium* are the pitiless portraits of ruined childhood. Living in poverty without present parental figures, using drugs, working or hustling instead of receiving a formal education, they have been deprived of the innocence of a normal childhood and of the opportunity to escape their condition of moral and material poverty.

An important difference between the post-war ruins portrayed in Rossellini's *Germania Anno Zero* and in *Disco Elysium* is that in the case of the latter, the Coalition military operation happened almost 50 years before the in-game events. While the ruins of Berlin have been either restored, replaced by new buildings or memorialised (De Martino 2011), Martinaise is purposely left in ruins by its occupier, not as a memorial but as a monition, a warning of what happens to those who oppose the Coalition and their neoliberal values. As a result, the aesthetic of Martinaise is one of precariousness. The roads are still damaged from the bombings, with a huge crater that has been repurposed as a playfield for pétanque. Pre-revolution buildings – bearing the traces of the Revolution and the Coalition military intervention – are patched up and inhabited, while cheap buildings and shacks have been built over the rubble. A drive toward reconstruction is however emerging, highlighting an ideological conflict between socialist and neoliberal planning strategies (see Casiello (ed.) 2011). These two approaches to reconstruction are embodied by two NPCs: the leader of the Dockworker's Union Evrart Claire and the Coalition Government official Charles Villedrouin. Evrart Claire embodies the socialist ideal of building anew, planning the construction of a youth center to replace a shantytown inhabited by fishermen and homeless alcoholics, a project undertaken for his own gain but also with the desire to improve the quality of life of the dockworkers. The appeal of re-urbanisation to capitalists is represented instead by Charles Villedrouin, encountered in the Capeside Apartment Complex, in the apartment of his young lover – a struggling homosexual art student – that he's allegedly renovating. Formerly a luxurious twelve-storey building with a beautiful sea view and a private dock, bombs and artillery fire heavily damaged the complex. The surviving lower floors – hastily repaired, with scaffolding in place to ensure their structural integrity – were then converted into cheap apartments, the sea view the only remnant of its past glory. Villedruin, admiring the view, is 'reminded of the former luxury of the place, discussing the potential market value of the apartments if they were to be restored.

While the district of Martinaise comprises the entirety of the game world, it is however inserted in a wider fictional world, with a detailed geographical and historical depth. Information about this wider fictional world is gathered by exploring and interacting with the game world and by talking with NPCs. Similarly to what was previously discussed regarding the focus in *Horizon Zero Dawn*, the interaction with the game world is mediated by an additional layer of textual information, conveyed as the player character's internal dialogue. Using a convention that is typical of the RPG genre, the game employs a skill system to allow players to customise their player character's abilities. Those skills not only determine the likelihood of success of specific actions in the game world – as dictated by the convention of the genre – but are also an integral part of the narrative design of the game. The skills are indeed metaphorised as Harry's subconscious voices, constructing a stream of consciousness that unfolds as the player explores and interacts with the game world. Like Aloy's focus, each skill reveals additional information, providing however not only specific bits of knowledge but its unique perspectives on the events. However, while Aloy's focus puts her in a condition of epistemic advantage, revealing information that is precluded to most NPCs, Harry's alcohol-induced amnesia caused him to forget even basic knowledge about the world – starting from his name and occupation – thus putting him in a situation of epistemic disadvantage towards his colleagues and the inhabitants of

Martinaise. As Vella and Cielecka (2021) argue, *Disco Elysium* uses the trope of the amnesiac protagonist not only to align the character's perspective with the player's lack of prior knowledge about the gameworld, but also to place the thematic focus on the reconstruction of the protagonist's past self "through fragmented memories or contextual evidence" (Vella/Cielecka 2021: 96). The murder investigation that the detective is tasked to solve with his partner Kim then becomes intertwined on one hand with Harry's investigation on his past self, and on the other hand with the discovery of the history and the mysteries of the fictional world. The ruination of Martinaise and its inhabitants is mirrored by the ruination of Harry's body and mind, devastated by alcohol and heartbreak, as both become the subject of an investigation that relies on the interpretation of fragments, whether memories or evidence. Deprived of the possibility to actually solve the murder – as the game introduces a culprit from outside the suspect pool, "meaning that the player has no real opportunity to reconstruct the hidden narrative of the crime for themselves before its true structure is revealed" (Novitz 2021: 34) – the player engages with the mysteries of the world and of Harry's mind in a polyphonic way, guided by the multiple perspectives of his skills. According to how players decide to distribute skill points, the corresponding voices will become stronger, revealing more information and providing different perspectives on the world (Novits 2021; Vella/Cielecka 2021). While Aloy's focus functions as a source of truth, providing her with a unifying perspective that helps her make sense of the ruination of the world and the imminent danger, Harry's voices instead further fragment and multiplicate the player's perception of the world, marking the impossibility of a positivist reading and a heroic resolution, thus shifting the game's focus from the achievement of both the player and the player character to the embodied exploration and discovery of the history and mystery of the game world.

In conclusion, the ruination represented in *Disco Elysium* does not only convey the history of the gameworld and its past and present conflicts, but is also part of a wider thematic focus on ruination in a broader sense, as both the biography of the player character and the fractured and polyphonic perspective of the internal discourse point towards ruination as an existential perspective. The neorealist portrayal of the tragedy of a district in ruins is thus muddled and complicated by multiple perspectives and registers – the political, the occult, the surreal – creating a narrative that is as fragmented as the ruins that it represents.

Conclusions

The analysis of the use of ruins as storytelling devices in the adventure games *Horizon Zero Dawn* and *Disco Elysium* revealed a series of commonalities both in regards to how ruins are employed and how players interact with them to discover the history of the game worlds. Firstly, both games employ ruins to convey the themes and mood of the game, echoing images of ruination in the real world that are deeply ingrained in the collective consciousness, in order to on one hand create a powerful emotional impact and on the other hand promote reflection over real world issues. Secondly, both games' usage of ruins moves beyond the visual backdrop, constructing explorable spaces that need active engagement from players to discover and interpret clues gathered from the ruins in order to reconstruct the history of the game world. Thirdly, in both games the ruins as visual clues are augmented by a layer of embedded narrative, storing textual and audio-visual information in specific locations in the game world waiting to be discovered by players. Despite both games employing a narrative structure based on indexical storytelling, requiring players to gather and interpret information, there's a substantial difference between the two regarding the knowledge resulting from this operation. In *Horizon Zero Dawn* the facts regarding the history of the game world are revealed through the discovery of testimonies of the past that progressively paint a clear picture of the events that led to the ecological apocalypse and their root causes. In *Disco Elysium*, instead, the information is gathered and presented in a conflicting way. Not only are the current and past issues of the fictional world not presented as black-and-white, with a plurality of forces and interests in play, but also the fragmentation of the inner monologue of the player character opens up to conflicting perspectives even upon seemingly banal facts. In other words, while a new unity emerges from the forgotten knowledge hidden among the ruins of the past apocalypse in *Horizon Zero Dawn*, providing the hero Aloy with all the information she needs to save the world, in *Disco Elysium* the very possibility of unity is forbidden, as Harry – a de-heroicized amnesiac alcoholic detective – navigates an all encompassing fragmentation that is impossible to reunify.

References

"Horizon Wiki: List of Real Locations in Horizon Zero Dawn" (n.d.). Retrieved from https://horizon.fandom.com/wiki/List_of_Real_Locations_in_Horizon_Zero_Dawn

"Horizon Wiki: Timeline" (n.d.). Retrieved from https://horizon.fandom.com/wiki/Timeline

"Oxford English Dictionary: Apocalypse" (n.d.). Retrieved from https://www.oed.com/dictionary/apocalypse_n?tl=true

Aghoro, N. (2022): "On Postapocalyptic Frontiers in *Horizon Zero Dawn*." In: Aghoro N./Filippaki I./Kempshall C./MacCallum-Stewart E./McCall J./Pöhlmann S. (eds), *Games and Spatiality in American Studies*. Berlin and Boston: de Gruyter, pp. 71-83.

Barros, A (2009): "Strategic Bombing and Restraint in 'Total War', 1915–1918." The Historical Journal 52(2), pp. 413-431.

Casiello, S. (ed.) (2011): *I Ruderi e la guerra. Memoria, ricostruzioni, restauri*. Firenze: Nardini.

Centeno, M. A./Enriquez, E. (2016): *War and Society*. Cambridge and Maiden: Polity.

Condis, M. (2020): "Sorry, wrong apocalypse: *Horizon Zero Dawn*, *Heaven's Vault*, and the ecocritical videogame." Game Studies, 20(3).

Craig, S. (2010): *Cinema After Fascism: The Shattered Screen*. Basingstoke and New York: Palgrave Mcmillan.

De Martino, G. (2011): "Ricostruzioni a Berlino." In S. Casiello (ed), I *Ruderi e la guerra. Memoria, ricostruzioni, restauri*. Firenze: Nardini, pp 33 – 51.

Domsch, S. (2019): "Space and narrative in computer games." In E. Aarseth/S. Günzel (eds), *Ludotopia: Spaces, Places and Territories in Computer Games*. Bielefield: transcript, pp. 103-123.

D'Orsi, A. (2007): *Guernica, 1937. Le bombe, la barbarie, la menzogna*. Roma: Donzelli.

Falkenhayner, N. (2021): "Futurity as an effect of playing *Horizon: Zero Dawn* (2017)" . Humanities, 10(2),72.

Fernández-Vara, C. (2011): "Game spaces speak volumes: Indexical storytelling." Proceedings of DiGRA 2011 Conference.

Fraser, E. (2019): "Post-apocalyptic Play: Representations of the End of the City in Video Games." In Trotta J. /Filipovic Z./Sadri H. (eds), *Broken Mirrors: Representations of Apocalypses and Dystopias in Popular Culture*. New York and London: Routledge, pp. 121-137.

Ginsberg, R. (2004): *The Aesthetics of Ruins*. Amsterdam – New York: Rodopi.

Grayling, A. C. (2006): *Among the Dead Cities: Is the Targeting of Civilians in War Ever Justified?* New York: Walker.

Heffernan, T. (2008): *Post-Apocalyptic Culture: Modernism, Postmodernism, and the Twentieth-Century Novel*. Toronto, Buffalo and London: University of Toronto Press.

Jelewska, A./Krawczak, M. (2018): "The Spectrality of Nuclear Catastrophe: The Case of Chernobyl." Politics of the Machines – Art and After. BCS Learning & Development.

Jenkins, H. (2004): "Game design as narrative architecture." Computer 44.3, pp. 118-130.

Jensen, L./Eaton, E. (2018): "From Guernica to Aleppo: the Price of Civilian Bombing in the Spanish Civil War." The Abraham Lincoln Brigade Archives.

Määttä, J. (2019): "The Future in Ruins: The Uses of Derelict Buildings and Monuments in Post-Apocalyptic Film and Literature." In Trotta J./Filipovic Z. /Sadri H. (eds), *Broken Mirrors: Representations of Apocalypses and Dystopias in Popular Culture*. New York and London: Routledge, pp. 138-156.

Novitz, J. (2021): "*Disco Elysium* as Gothic Fiction." *Baltic Screen Media Review*, (9), pp. 32-42.

Pagels, E. (2012): *Revelations: Visions, Prophecy, and Politics in the Book of Revelation*. New York: Viking Penguin.

Rondolino, G. (1989): *Roberto Rossellini*. Torino: UTET.

Rush-Cooper, N. (2020): "Nuclear Landscape: Tourism, Embodiment and Exposure in the Chernobyl Zone." Cultural Geographies, 27(2), pp 217-235.

Şengün, S. (2017): "Ludic Voyeurism and Passive Spectatorship in *Gone Home* and Other Walking Simulators." *Video Game Art Reader*, 1, pp 30-42.

Steinhauser, G./Brandl, A./Johnson, T. E. (2014): "Comparison of the Chernobyl and Fukushima Nuclear Accidents: A Review of the Environmental Impacts." *Science of the Total Environment*, 470, pp. 800-817.

Stone, P. R. (2013): "Dark Tourism, Heterotopias and Post-Apocalyptic Places: The Case of Chernobyl." In White L./Frew E. (eds), *Dark Tourism and Place Identity: Managing and Interpreting Dark Places*. Abingdon and New York: Routledge, pp. 79-93.

Vella, D. (2010): *Virtually in Ruins: The Imagery and Spaces of Ruin in Digital Games*. Masters Dissertation. Malta: University of Malta.

Vella, D/Cielecka, M. (2021): " 'You Won't Even Know Who You Are Anymore': Bakthinian Polyphony and the Challenge to the Ludic Subject in *Disco Elysium*." *Baltic Screen Media Review*, 9(1), pp. 90-104.

Weisman, A. (2007): *The World Without Us*. New York: St. Martin's.

List of Artworks and Games

Darwell, John (1998): Legacy: Inside the Chernobyl Exclusion Zone.
Fisher, Nina/el Sani, Maroan (2021): Contaminated Home.
Guerrilla Games (2017): Horizon Zero Dawn.
Rossellini, Roberto (1948): Germania Anno Zero.
ZA/UM (2019): Disco Elysium – The Final Cut.

Biographical Notes

Romi Sofia Abatangelo is an early career scholar, indie game designer and amateur performer. Their research interest and artistic production move between theatre and games, with a focus on the narrative potential of games and the ludic traits of performing arts. Abatangelo received their M.Sc. from the Institute of Digital Games at the University of Malta, with a thesis focused on the spatial, narrative and participatory qualities of walking simulators and explorative immersive theatre.

Rahul Bishnoi is a PhD Student in Theatre and Performance Studies at Tufts University. His research interest includes exploring theatre as a research method to study the performativity of spaces. He is also a Global Majority and Underrepresented Writing Scholar at the Oxford Centre for Life Writing and a visiting artist at Maison Julien Gracq in 2025.

Trevor Borg is Associate Professor and Head of the Department of Digital Arts at the University of Malta, as well as a practicing artist. His work has been presented at major international exhibitions including the 58th edition of the Venice Biennale (2019), the European Parliament (2012), the Beijing International Art Biennale (2022), the Malta Biennale (2024), and in numerous other countries. He holds a PhD in Fine Art from the University of Leeds. He is a member of the creative practice-led research network LAND2 and an associate at Cut Contemporary Fine Arts Lab at the Cyprus University of Technology.

Mathias Fuchs is an artist, musician and media scholar. He was Principal Investigator for a German Research Council funded project on Gamification (2018-2021) at the Institute of Culture and Aesthetics of Digital Media (ICAM) at Leuphana University, Lüneburg. He was a Senior Fellow at International Research Center for Cultural Studies (IFK) Linz and teaches currently at Leuphana University, Lüneburg and at the University of Applied Arts in Vienna. His monograph *Phantasmal Spaces* has been published by Bloomsbury London, New York.

James Manning is a videogame researcher, creator, and Lecturer at RMIT University in Naarm/Melbourne, Australia. His research investigates videogame design, digital materiality, media fandom, play cultures, and digital cultural heritage. Recent publications include articles on videogame design, livestreaming, cultural memory, and archiving play practices. He is Co-Chair at *Freeplay*, the world's longest-running independent games festival, and a Managing Editor for the *Journal of Games Criticism.*

Lawrence May is a Senior Lecturer at the University of Auckland, New Zealand. His research explores meaning-making in player communities and the entanglement between videogames and ecological crisis, and has appeared in journals including *Game Studies* and *Games and Culture*. Lawrence is the author of *Digital Zombies, Undead Stories* (Bloomsbury, 2021), which examines emergent narrative in multiplayer contexts and the place of zombies in contemporary videogames.

Daniele Monaco is a PhD student in the philosophy of dwelling at the University of Perugia, Italy, researching the relationship between community and place through the hermeneutic lens of *Genius Loci*. His work explores the ontological foundations of our connection to places and examines how this bond evolves across both physical and digital environments. His current focus is on how virtual spaces—particularly video games—serve as new arenas for dwelling, reshaping our understanding of presence, belonging, and meaning. By bridging Heidegger's philosophy of dwelling with interactive digital narratives, his research seeks insights into how we inhabit and interpret space across both real and virtual domains.

Souvik Mukherjee is Associate Professor in Cultural Studies at the Centre for Studies in Social Sciences Calcutta, India. Souvik is the author of three monographs, *Videogames and Storytelling: Reading Games and Playing Books* (Palgrave Macmillan 2015), *Videogames and Postcolonialism: Empire Plays Back* (Springer UK 2017) and *Videogames in the Indian Subcontinent: Development, Culture(s) and Representations* (Bloomsbury India 2022. Souvik has been named a Digital Games Research Association (DiGRA) Distinguished Scholar in 2019 and a Higher Education Video Game Alliance (HEVGA) fellow in 2022. He is also an affiliated senior research fellow at the Centre of Excellence, Game Studies at the University of Tampere.

Caio Tulio Olimpio Pereira da Costa is an Adjunct Professor of Digital Games at the State University of Bahia (UNEB). PhD in Technological Education and M.A. in Communication, both from the Federal University of Pernambuco (UFPE, 2024; 2020). He worked as a Visiting Scholar/Research Intern at the University of Quebec in Outaouais (UQO, 2024) and is a member of the Digital Media and Intercultural Mediation Research Group at the Federal University of Pernambuco (UFPE) and of CIPEG – Interdisciplinary Research Collective in Games. His research interests include Cyberculture, Digital Communication, Game Studies, Technologies, Image and Imaginary, Narratives, Languages, Technological Education, Cinema, and Pop Culture.

Michael Stock is Visiting Assistant Professor of Film Studies and English at Pepperdine University and Adjunct Professor of Cinema and Media Studies at Southern California Institute of Architecture [SCI-Arc]. His research interests in critical theory and the histories of film, television, transmedia, and special effects are reflected in his extensive published work and the wide array of courses he teaches. Stock received his Ph.D. in Cinema and Media Studies at UCLA, and his M.A. and B.A. in English at the University of Nebraska, Lincoln.

Ian Sturrock is a game designer, with a variety of tabletop RPG publications under his belt including games, supplements, and adventures for Pelgrane Press, Green Ronin, Hogshead Publishing, Modiphius, Guardians of Order, and Mongoose Publishing. In the digital games sphere he's worked for Sony, Ubisoft, and others. Ian teaches and researches at Teesside University, with particular interests in motivation for play, game design, game studies, tabletop RPGs, and games narrative. In 2024 he spent three months at Tampere University as a Visiting Scholar. In his spare time, he wargames, larps, plays capoeira, and runs a small press tabletop RPG publisher, Serpent King Games.

Ana Laura Torquato is an artist, researcher and PhD Candidate in Art and Visual Culture at the Federal University of Goiás (UFG), focusing on Artistic Poetics and Creation Processes. She also served as a visiting scholar at the University of Malta (UM) in 2024, supported by the Coordination for the Improvement of Higher Education Personnel (CAPES). She is a member of the Ciberart and Creation Research Group at the Federal University of Goiás (UFG) and of CIPEG – Interdisciplinary Research Collective in Games. Her research interests focus on Horror Games, Visualities, Technologies and Game Studies within the framework of Visual Culture.

Daniel Vella is the Director of the Institute of Digital Games at the University of Malta. He is the co-author of *Virtual Existentialism: Meaning and Subjectivity in Virtual Worlds* (Palgrave 2020 - with Stefano Gualeni), and has published his research in edited volumes and journals including Game Studies, Countertext, ToDiGRA, Eludamos, and Techné: Research in Philosophy and Technology. He also works as a writer and narrative designer for games, most recently on *Fateforge: Chronicles of Kaan* (Mighty Boards 2024). His debut novel is due to be published by Praspar Press in 2026.

Timothy J. Welsh is an Associate Professor of English at Loyola University New Orleans. He is the author of *Mixed Realism: Videogames and the Violence of Fiction* (University of Minnesota Press, 2016) and co-author of *Video Games, Literature, and Close Playing A Practical Guide* (Routledge, 2025). He has published several articles on video games, interactive narrative, and digital society and serves as managing editor of the *Journal of Games Criticism*.